D1485655

Using Remote Sensing in State and Local Government
Information for Management and Decision Making

Steering Committee on Space Applications and Commercialization

Space Studies Board
Division on Engineering and Physical Sciences

NATIONAL RESEARCH COUNCIL
OF THE NATIONAL ACADEMIES

THE NATIONAL ACADEMIES PRESS
Washington, D.C.
www.nap.edu

THE NATIONAL ACADEMIES PRESS   500 Fifth Street, N.W.   Washington, DC 20001

NOTICE: The project that is the subject of this report was approved by the Governing Board of the National Research Council, whose members are drawn from the councils of the National Academy of Sciences, the National Academy of Engineering, and the Institute of Medicine. The members of the committee responsible for the report were chosen for their special competences and with regard for appropriate balance.

Support for this project was provided by National Aeronautics and Space Administration Contracts Nos. NASW-96013 and 01001, National Oceanic and Atmospheric Administration Contract No. 50-DKNA-6-90040, Stennis Space Center Order Nos. NS-7426 and 7570, Environmental Protection Agency Grant No. X-82821401, Department of Transportation Order No. DTRS56-00-P-70077, U.S. Geological Survey Cooperative Agreement No. 00HQAG0204, and Department of the Army Order No. DACA89-99-M-0147. Any opinions, findings, conclusions, or recommendations expressed in this material are those of the authors and do not necessarily reflect the views of the sponsors.

The cover was designed by Penny Margolskee.

Cover images (left to right):

1. Landsat 7 Enhanced Thematic Mapper Plus image, New York, April 14, 2001.
2. QuickBird, 60 cm, pansharpened satellite image of the Brooklyn Bridge as it enters Manhattan. Image taken August 2002. Courtesy of DigitalGlobe.
3. U.S. Geological Survey aerial photograph, Lower Manhattan, April 10, 1997.

International Standard Book Number 0-309-08863-1

Copies of this report are available free of charge from:

Space Studies Board
National Research Council
500 Fifth Street, N.W.
Washington, DC 20001

THE NATIONAL ACADEMIES
Advisers to the Nation on Science, Engineering, and Medicine

The **National Academy of Sciences** is a private, nonprofit, self-perpetuating society of distinguished scholars engaged in scientific and engineering research, dedicated to the furtherance of science and technology and to their use for the general welfare. Upon the authority of the charter granted to it by the Congress in 1863, the Academy has a mandate that requires it to advise the federal government on scientific and technical matters. Dr. Bruce M. Alberts is president of the National Academy of Sciences.

The **National Academy of Engineering** was established in 1964, under the charter of the National Academy of Sciences, as a parallel organization of outstanding engineers. It is autonomous in its administration and in the selection of its members, sharing with the National Academy of Sciences the responsibility for advising the federal government. The National Academy of Engineering also sponsors engineering programs aimed at meeting national needs, encourages education and research, and recognizes the superior achievements of engineers. Dr. Wm. A. Wulf is president of the National Academy of Engineering.

The **Institute of Medicine** was established in 1970 by the National Academy of Sciences to secure the services of eminent members of appropriate professions in the examination of policy matters pertaining to the health of the public. The Institute acts under the responsibility given to the National Academy of Sciences by its congressional charter to be an adviser to the federal government and, upon its own initiative, to identify issues of medical care, research, and education. Dr. Harvey V. Fineberg is president of the Institute of Medicine.

The **National Research Council** was organized by the National Academy of Sciences in 1916 to associate the broad community of science and technology with the Academy's purposes of furthering knowledge and advising the federal government. Functioning in accordance with general policies determined by the Academy, the Council has become the principal operating agency of both the National Academy of Sciences and the National Academy of Engineering in providing services to the government, the public, and the scientific and engineering communities. The Council is administered jointly by both Academies and the Institute of Medicine. Dr. Bruce M. Alberts and Dr. Wm. A. Wulf are chair and vice chair, respectively, of the National Research Council.

www.national-academies.org

Preface

For several years, the Space Studies Board has recognized that because of recent advances in the spatial, spectral, and temporal resolution of available data, there are more opportunities for developing practical applications of remote sensing data. The combination of technological advances in remote sensing; widely available, compatible geographic information technologies; increased availability of data at usable scales; and greater diversity in data sources and infrastructure support has made widespread and diverse applications possible in a broad variety of new sectors. At the same time, however, changes have been taking place in the roles played by data producers and consumers in the public and private sectors, the universities, and the value-added community. Changes in the economic and policy environment for remote sensing, the growth of a commercial remote sensing industry, the expansion and proliferation of data sources worldwide, and the greater breadth of remote sensing data provided by federal agencies are the result of interacting market, technological, policy, and budgetary opportunities and incentives.

To explore the implications of these significant changes in the environment for the production and use of remote sensing data and information, the Space Studies Board appointed the Steering Committee on Space Applications and Commercialization and initiated a series of three workshops. Each workshop dealt with a different area: (1) the extension of remote sensing technologies and products into operational applications through technology transfer; (2) the impact of the new and evolving remote sensing environment on basic research in the Earth sciences; and (3) the development and use of remote sensing applications in the public sector, specifically state and local government. The board obtained sponsorship for the workshops from several government agencies: the National Aeronautics and Space Administration (Headquarters and Stennis Space Center), the National Oceanic and Atmospheric Administration (National Environmental Satellite, Data, and Information Service and National Ocean Service), the Environmental Protection Agency, the Department of Transportation, the U.S. Army Corps of Engineers, and the U.S. Geological Survey.

The first workshop, "Moving Remote Sensing from Research to Applications: Case Studies of the Knowledge Transfer Process," was held in Washington, D.C., in May 2000 and resulted in the report *Transforming Remote Sensing Data into Information and Applications* (National Academy Press, Washington, D.C., 2001). The second workshop, held in Washington, D.C., in March 2001, was titled "Remote Sensing and Basic Research: The Changing Environment." It, too, resulted in a report, *Toward New Partnerships in Remote Sensing: Government, the Private Sector, and Earth Science Research* (National Academy Press, Washington, D.C., 2002).

This is a report on the third workshop, "Facilitating Public Sector Uses of Remote Sensing Data," held in January 2002. Its focus was on the development and use of remote sensing data and information by state, local, and regional governments. The workshop was attended by public sector officials from state, local, regional, and federal agencies as well as remote sensing data providers (both satellite and airborne) from the public and private sectors and university scientists. To encourage the participation of representatives of the nonfederal public sector, the workshop was held not in Washington, D.C., but in Boulder, Colorado. Like the previous two workshops, this one was preceded by a planning meeting to which representatives of the sponsoring agencies and other interested individuals were invited.

The charge to the steering committee was to examine the opportunities and the potential challenges and policy issues associated with the application and use of remote sensing data in the public sector. Specifically, the steering committee was asked to

• Identify opportunities, approaches, bottlenecks, and procedures for government agencies to use, apply, and share remote sensing data with other institutions;

• Provide a forum for government science and operational agencies and policy and commercial entities to exchange information, priorities, experience, and procedures on remote sensing applications; and

• Identify barriers to the development and use of remote sensing applications across the government.

In preparing this report, the steering committee (Appendix B) drew on information from several sources: the workshop itself, including presentations, written submissions by participants, and discussions (see the workshop agenda in Appendix C); information presented at the planning meeting in August 2001; and steering committee research and deliberations during and after the workshop. The steering committee also drew heavily on its first report, *Transforming Remote Sensing Data into Information and Applications,* because the issues discussed in that report also apply to state and local governments (see Appendix A for the Executive Summary of that report). Although it is not a formal study of public sector remote sensing applications, this third report in the series reflects input to the steering committee from a broad spectrum of public officials and data providers.

This report is addressed primarily to nontechnical managers and decision makers in state, local, and regional governments. It both introduces them to the subject of remote sensing applications and raises significant institutional, budgetary, legal, intergovernmental, and technical issues related to the development of effective, operational remote sensing applications in the public sector. The steering committee decided to direct the report primarily to managers, because they have decision-making authority on remote sensing and other technological resources but may have fewer opportunities to understand the technology and its implications than do staff in state and local government who use geospatial information. The report is also directed at those who have responsibility for geospatial data and information in the public sector, both federal and nonfederal, and to remote sensing data providers in both the public and private sectors who may benefit from further insight on the particular challenges of using remote sensing data in state and local government.

Acknowledgment of Reviewers

This report has been reviewed in draft form by individuals chosen for their diverse perspectives and technical expertise, in accordance with procedures approved by the National Research Council's (NRC's) Report Review Committee. The purpose of this independent review is to provide candid and critical comments that will assist the institution in making its published report as sound as possible and to ensure that the report meets institutional standards for objectivity, evidence, and responsiveness to the study charge. The review comments and draft manuscript remain confidential to protect the integrity of the deliberative process. We wish to thank the following individuals for their review of this report:

Eric A. Anderson, City of Des Moines, Iowa;
Patrick J. Bresnahan, Richland County, South Carolina;
Joseph Engeln, Missouri Department of Natural Resources;
Donald Rundquist, University of Nebraska, Lincoln; and
Jan Svejkovsky, Ocean Imaging, Inc.

Although the reviewers listed above have provided many constructive comments and suggestions, they were not asked to endorse the conclusions or recommendations, nor did they see the final draft of the report before its release. The review of this report was overseen by John V. Evans, Comsat Corporation (retired). Appointed by the National Research Council, he was responsible for making certain that an independent examination of this report was carried out in accordance with institutional procedures and that all review comments were carefully considered. Responsibility for the final content of this report rests entirely with the authoring committee and the institution.

Contents

APPENDIXES

Executive Summary

The past decade has seen significant improvement in the spatial and spectral resolution of the civil remote sensing data available to state, local, and regional governments. With the development of advanced airborne remote sensing technologies like lidar (light detection and ranging) and the launch of high-resolution, commercial remote sensing satellites, state and local jurisdictions now have the opportunity to obtain digital data at resolutions approaching those of aerial photography. State and local users of remote sensing data can also access data from the Landsat series for comparisons and detection of change over decades.

As important as these improvements in the quality and availability of remote sensing data is the growing number of geospatial data management and analysis tools available for use by state and local governments. With geographic information systems (GIS), digital remote sensing data can now be integrated with other types of digital data currently managed by state and local governments. Such technological advances can foster the development of remote sensing applications in the nonfederal public sector.

However, the use of remote sensing data and applications involves more than the underlying technical capacity. From the perspective of the remote sensing applications end user in state, local, or regional government, what is important is the information that remote sensing applications can make available, not the raw data per se. Equally important, the ability of a state or local government agency or jurisdiction to take advantage of recent technological advances depends on institutional, leadership, budgetary, procedural, and even personnel factors.

To examine the full range of factors that have led to the development of successful applications of remote sensing data in state and local governments and to identify common problems encountered in this process, the Space Studies Board's Steering Committee on Space Applications and Commercialization organized the workshop "Facilitating Public Sector Uses of Remote Sensing Data." Presentations at the workshop included case studies of the adoption and use of remote sensing applications in local government (Baltimore, Maryland; Richland County, South Carolina; and Boulder County, Colorado), state government (Missouri, Washington, and North Carolina), and regional government (the Portland metropolitan area in Oregon and the communities of the Red River Valley along the North Dakota–Minnesota boundary); information on remote sensing applications in specific sectors and on patterns of adoption; and technical material on sensors. The workshop was attended by representatives of state, local, and regional governments, the federal government, the private sector, and universities.

The case studies illustrate some, not all, of the uses of remote sensing data in state and local government. The issues they raise are not specific to any single type of data or application. At the same time, certain uses of remote sensing data, especially in operational applications, may involve challenges and issues that are not directly addressed in this report.

This report draws on the information presented in the workshop, the workshop planning meeting with agency sponsors, and the expertise and viewpoints of the steering committee. For this reason, technical information is kept to a minimum. The report and its recommendations are the consensus of the steering committee and not necessarily of the workshop participants. The report is directed to those in state, local, and regional governments who make crucial decisions about both the commitment of resources to developing remote sensing capabilities and the use of remote sensing information in the public sector. The steering committee envisions that the report will also be useful to geospatial professionals in state, local, and regional government who work with those managers and decision makers; to remote sensing data providers in the federal government and the private sector; and to federal officials who interact with the nonfederal public sector on issues that require geospatial data.

BASIC OBSERVATIONS

The steering committee found that the context for the use of remote sensing applications in state, local, and regional governments differs significantly from that for other applications. Responsible primarily for providing public services and governance, state and local governments are supported by tax revenues that can vary considerably from year to year, and political considerations can also influence decision making. Many workshop participants spoke of budget shortfalls and stringencies they were experiencing or expected to experience in the next several years that could negatively influence the adoption and use of remote sensing data and information.

From the perspective of a commercial firm seeking to supply remote sensing data or services, the scale of the public sector can pose serious problems. The sheer number of state, local, and regional governments increases the costs of providing them with remote sensing data. There are 50 states and more than 3,100 counties in the United States; the New York City metropolitan area alone contains 31 counties and over 1,600 jurisdictions. In addition, nonfederal government decision making about new technologies is complicated and often requires buy-in from multiple parts of the government. Even if there were some timely way to determine which jurisdictions were prepared to buy remote sensing data or services, negotiating separate small contracts might not be cost-effective for large, commercial remote sensing firms.

From the perspective of state and local governments, moreover, there are benefits of working with local firms and universities rather than with data or other service providers from outside the immediate region. Proximity has always been a factor for governments working with small aerial photography firms, for example, and local firms establish long-term relationships with local government agencies.

The steering committee found that adoption of remote sensing data and information products in the nonfederal public sector has been affected by several aspects of policy and operations. These include (1) financial and budgetary constraints; (2) institutional, organizational, and political issues; (3) the geospatial experience, skills, and training available in the jurisdiction; (4) the capacity to make the transition from photographic to digital data; and (5) licensing and data management. The steering committee also found that the adoption of remote sensing data and applications is often related to having a strong advocate for the new technology who can persuade technical personnel, managers, decision makers, and even the public about the utility of the data and information.

FINDINGS AND RECOMMENDATIONS

Improving Management and Efficiency

It is advantageous for public sector jurisdictions considering the use of new remote sensing technologies to learn from the organizational practices of governments that have already used remote sensing applications successfully.

Geospatial Data Management

Finding: Some state and local governments have taken an ad hoc, decentralized approach to using remote sensing data. Individual departments or offices took it upon themselves to obtain the remote sensing data they needed for a specific application or project. Where there was no city- or statewide inventory of data, the independent purchases of data resulted in multiple acquisitions of the same remote sensing images and inefficient management and use of geospatial data resources. Certain municipal governments, however, took a more centralized approach, locating remote sensing resources within geospatial data or information offices under the direction of technical staff proficient in the use of geospatial data.

Budgetary and staffing limitations, coupled with the increased convergence of digital technologies, including geospatial data from GIS, satellite, and airborne remote sensing and even global positioning systems, suggest that an approach in which a single administrative entity manages geospatial data is more cost-effective than a decentralized approach and facilitates use of the data by state and local governments.

Recommendation 1: A state, local, or regional government should consider making a single unit responsible for managing its geospatial data, information, and technologies.

Cross-Jurisdictional Remote Sensing Data Cooperatives

Finding: The cost of obtaining and managing remote sensing data can be prohibitive for state, local, and regional government departments or agencies, particularly during a period of budgetary shortfalls. The steering committee found that some governments in the nonfederal public sector have successfully joined together to form local or regional cooperatives or consortia that purchase remote sensing data for all members of the group. Data cooperatives can also help small jurisdictions to manage remote sensing and other digital data.

Recommendation 2: Public officials responsible for obtaining and using geospatial data should examine the benefits of forming multijurisdictional consortia or cooperatives to reduce duplication of cost and effort.

Procurement Processes

Finding: Public sector procurement processes for the purchase of remote sensing data can be lengthy and time-consuming, making it difficult for a jurisdiction to obtain timely authorization for purchasing such data. In addition, public sector accounting processes are most effective in dealing with marginal changes in budgets that are relatively constant from year to year. Remote sensing data may constitute a major purchase needed on an irregular basis, which can be difficult to accommodate in normal public sector accounting practices.

Recommendation 3: State and local government budget and procurement practices should be examined and modified, if necessary, to facilitate acquisition of multiyear remote sensing data.

An independent body such as the Government Accounting Standards Board—a private, nonprofit institution that develops accounting reporting standards for state, local, county, and other nonfederal government entities—or another independent accounting organization could be consulted for input on how to account more effectively for expenditures on remote sensing data.

Recommendation 4: State and local governments should explore the feasibility of establishing long-term purchase agreements with local institutions or vendors to give themselves flexibility in obtaining remote sensing data.

Creating a More Effective Public Sector Market for Remote Sensing Data

A large and active public sector market for remote sensing data and information will provide economies of scale for governments seeking cost-effective remote sensing applications and for the public, private, and international vendors that supply data and services to state and local governments (see "Working with the Private Sector," in Chapter 4). The steering committee learned several ways in which a more active and effective market for state and local applications of remote sensing data and information can be created.

Standards for Digital Spatial Data and Information Products

Finding: The increasing use of digital remote sensing data rather than photographic data by state and local governments means that new standards are needed for digital spatial data and information. The advantages of commonly accepted digital spatial data standards include reduced cost, improved ability to use the data for multiple purposes, standardization of technical training, and quality assurance. The adoption of digital data standards would require that procurement regulations for many state and local government entities be revised. Common standards for digital data could be developed by a coordinating body funded by the federal government that includes representatives of both data users and data providers. The federal agencies involved in the effort could determine which agency should take the lead.

Recommendation 5: The U.S. government, in collaboration with professional organizations, state and local governments, and vendors, should take the lead in establishing standards for digital spatial data and information products.

Private Actions to Build a Public Sector Market

Finding: Although commercial providers of remote sensing data recognize the potential economic significance of the nonfederal public sector market for remote sensing data, they often do not do enough to stimulate its development and growth.

Recommendation 6: To help remedy the lack of trained remote sensing personnel in state and local governments and to raise awareness of the advantages of working with satellite remote sensing data, commercial satellite data providers and remote sensing digital image processing vendors should look to GIS software companies as models for building the state and local government market.

Licensing

Finding: The licensing provisions of commercial satellite data companies seem restrictive, offering little flexibility to state and local governments. Strictly followed, commercial licensing provisions can add to the cost of data in the nonfederal public sector and can result in redundant purchases of the same data within a single jurisdiction, creating a disincentive for state, local, and regional governments to purchase data from the private sector. Although representatives of private remote sensing firms suggested that it is possible to negotiate new licensing agreements based on specific needs, officials in the nonfederal public sector reported that they had not been made aware of this flexibility.

Recommendation 7: Private sector providers of remote sensing images should offer standard information about flexibility in their pricing policies, ensuring that the information is widely available, especially information about establishing jurisdiction-wide site licenses or long-term purchase agreements for state and local governments.

Opportunities to Work with the Public Sector

Finding: There is no single source of information on prospective remote sensing data needs of state, local, and regional governments. This limits the market to local firms or those that have personal contacts with a jurisdiction seeking bids for data or services. The failure to notify a larger potential contractor community may stifle competition and result in higher costs.

Recommendation 8: Associations of state and local governments should establish national or statewide opportunities/forums for state, local, and regional governments to advertise their needs for remote sensing data.

Cooperation Between the Federal and Nonfederal Public Sectors

Finding: The steering committee found widespread cooperation between federal agencies and state, local, and regional governments in initiating remote sensing applications programs. Much of this cooperation, however, took place within federal programs that support state and local government use of remote sensing data for specific programmatic objectives. Some state and local government representatives are seeking general infrastructure, support, or guidance on how they might take advantage of remote sensing data or applications programs supported by the federal government. There appears to be an unfulfilled need for a point of contact at federal agencies to help state and local users obtain information and facilitate collaboration between state and local users and federal agencies.

Recommendation 9: Federal agencies should have a formal point of contact for representatives of state and local governments that need technical assistance or want to identify sources of financial assistance for their use of remote sensing applications.

1

Introduction

The growing interest in using remote sensing data and information in public sector management and decision making is being fed by the perceived utility of remote sensing data and information for a variety of state and local government purposes, by the increasing availability and usability of remote sensing-derived data in geographic information systems (GIS) and decision support systems, and by improvements in the spatial and spectral resolution of the data (see Box 1.1 for definitions of key terms). Since the early days of aerial photography in the 1930s and 1940s, local governments have been among the most active users of information obtained from remote sensing. State and local agencies have long depended on remote sensing technology for interpreting the aerial photographs they use to map and monitor changes in, for instance, land use, civil (nonmilitary) infrastructure, and transportation. The advent of civil satellite remote sensing raised expectations that state, county, and local governments would rapidly develop further applications of this new source of information, but those early expectations were not fulfilled. Although many state and local governments have used airborne remote sensing data to obtain information for both management and policy purposes, the use of satellite data in the nonfederal public sector is still limited.

In some jurisdictions and for some purposes, airborne remote sensing continues to be preferred over the use of satellite data because of the former's high spatial resolution and stereoscopic features. High-resolution images are necessary for many urban uses, and the resolution of the early civil satellite data was too low. By the end of the 1990s, however, new sources—federal and commercial—of high-resolution and multispectral remote sensing data from satellites led

BOX 1.1
Definitions of Key Terms

Understanding the advantages and potential uses of remote sensing begins with an understanding of key terms, some of which are defined here as they apply to the more common aspects of remote sensing.

Hyperspectral data may be collected in hundreds of wavelengths and therefore offer the richest detail on the physical or biophysical properties of the objects being viewed.

Light detection and ranging (lidar) measures the time it takes for a pulse of laser energy to travel round-trip from the laser source to a target and back to a sensor on the same satellite (or airplane) as the laser source. The pulse's travel times are useful for measuring the elevation of a hard surface such as a building or the ground. Newer systems can collect data that distinguish reflections from multiple sources such as trees and the ground surface below them and can thus provide information on the structure of a forest.

Multispectral data include information detected in several electromagnetic wavelengths and thus provide more detail on the observed objects, as opposed to black and white (or panchromatic) data, which do not provide any spectral information.

Orthophotography refers to aerial photographs that are referenced to precise *x, y* coordinates on the ground. Referencing removes from the photographs the distortions caused by the terrain, Earth's curvature, and camera angle. Orthophotography creates images that have the characteristics of a map: One can measure distances, areas, and angles accurately. These features make orthophotography useful for urban planning, environmental assessment, and other applications.

Photogrammetry is the art, science, and technology of obtaining reliable information about physical objects and the environment through processes of recording, measuring, and interpreting photographic images, generally taken from an aircraft.

Remote sensing is a means of obtaining data and images from sensors or cameras located at a distance rather than from direct human observation. Remote sensing data can be collected in several ways. Aerial images are obtained by photographing Earth's surface from an airplane. Sensors on satellites generally provide digital rather than photographic images; they measure the electromagnetic radiation reflected or emitted from vegetation and terrain, which is then converted into the necessary information.

Spatial resolution refers to the smallest feature discernable in an image. For example, a single picture element (pixel) in an image collected from the Landsat satellite measures 30 meters on a side, while some commercial satellites collect images that can distinguish features as small as 0.5 meter. While one could not distinguish an automobile using data with 30-meter resolution one could do so using data with 0.5-meter resolution.

(continues)

Spectral resolution refers to a sensor's ability to collect data at specific electromagnetic wavelength ranges. Higher resolution provides more information about the physical characteristics of the object observed or more details about the biophysical properties of land cover and plants.

Temporal resolution refers to the frequency with which a satellite revisits the same location. More frequent observations provide a better record of change, such as plants greening up in the spring, and increase the chance of observing a short-term event. For example, the Landsat 7 satellite has a temporal resolution of 16 days.

many to conclude that state and local government would now become a significant, possibly the most significant, user of the data.

Another type of remote sensing is ground-based and consists of sensors on Earth's surface such as air-pollution sensors, ground-penetrating radar, and sonar, which are used to obtain information about Earth. Ground-based remote sensing is not, however, covered in this report, which explores the use of new as well as older types of satellite and airborne remote sensing data and information.

This report explores how state, local, and regional governments apply remote sensing data and information applications, the problems they have met with in these applications, and how these problems are being addressed.[1] The report is intended as both an introduction to remote sensing applications and a guide for managers and decision makers in state and local government who are responsible for supervising, establishing, managing, or budgeting for public sector data and information, including geospatial data operations. It is also intended for technically sophisticated geospatial professionals in state and local government who are interested in remote sensing applications and technology and for decision makers in the federal and commercial sectors who are engaged in the production, regulation, or dissemination of civil remote sensing data, to help them understand the opportunities offered by satellite remote sensing data in the nonfederal public sector as well as the constraints such data pose.

The nonfederal public sector in the United States is a unique and often difficult setting in which to introduce new applications of satellite remote sensing data. Although potentially a large user market in the aggregate, it is highly decentralized, consisting of tens of thousands of independent and quasi-independent jurisdictions, each with complex budgetary, procurement, and decision-

---

[1]Another NRC report, *People and Pixels: Linking Remote Sensing and Social Science*, Washington, D.C., National Academy Press, 1998, discusses the use of remote sensing data by social scientists, farmers, local governments, and urban and natural resource managers.

making processes that influence the acquisition of remote sensing imagery. These public sector jurisdictions operate under significant budget constraints, and the budgets can change from year to year. Because managers in these jurisdictions are directly accountable to the voting public and their elected representatives and often must also meet the requirements of other levels of government, such as states and federal agencies, they operate in a complex political environment.

Unlike the federal government, which uses remote sensing data and information for research, analysis, and public policy, state and local governments are engaged primarily in fulfilling highly specific operational responsibilities related to public sector management and governance. They obtain the data and information required for those responsibilities by traditional information-gathering procedures that may be legally constrained and labor-intensive. In such a setting, introducing new, high-technology data sources like satellite remote sensing can be difficult and expensive from technological, administrative, and budgetary perspectives. As a result, public sector managers may believe that adopting remote sensing data from satellites creates more problems than it solves.

To assess the potential value of satellite remote sensing data to their operations, managers in state and local government need some understanding of technical distinctions and of the capabilities of remote sensing and its applications. They also need to understand the institutional and budgetary ramifications of adopting satellite remote sensing applications in a public sector setting. This report, which is based on information supplied in public sector case studies, attempts to meet the need for both types of information.

At the same time, the steering committee recognizes that public sector decision makers who are not technical experts will and should turn to specialists for an assessment of technical issues. For this reason, and because the technology frequently changes, technical information in this report is kept to a minimum. Table 1.1 briefly lists types of remote sensing data that may be useful in state and local government operations. The steering committee hopes in this way to introduce nontechnical decision makers to the vocabulary of remote sensing and its capabilities. The focus of the report, however, is the utility of remote sensing applications and the institutional, budgetary, and policy issues that arise when remote sensing data and information are used to meet state and local government information needs.

Many of the examples and case studies presented describe land remote sensing applications because of the broad relevance and importance of land data to state and local governments. The issues raised in the case studies are not specific to any single data type or application; moreover, specific uses of remote sensing data may introduce issues other than those discussed in this report. Although the emphasis is on civil remote sensing, the report recognizes that since September 11, the lines between civil and national security data in state and local government have blurred.

This report grew out of the workshop "Facilitating Public Sector Uses of

TABLE 1.1 Selected Types of Remote Sensing Data

Data Type	Attributes	Availability	Case Studies Supplying Examples
High-resolution optical	—High detail within small area —Useful for infrastructure, urban applications —Provides high-resolution views of objects at 1-2 m	—Private U.S. and non-U.S. remote sensing companies —Private aerial photography firms	—Baltimore, Md. —Richland County, S.C. —Washington State
Multispectral optical, medium resolution	—Shows multiple geophysical features —Allows for classification of vegetation, land use —Provides medium-resolution views of objects at 10-30 m —Beneficial for viewing large land areas	—U.S. and non-U.S. government agencies and private firms	—Baltimore, Md. —Portland (Ore.) Metro —Boulder County, Colo. —Missouri —Richland County, S.C.
Hyperspectral	—Provides maximum information on objects viewed —Allows for high-level classification, identification, and distinguishing of objects	—U.S. and non-U.S. remote sensing companies	—These data may be useful for identifying roofing materials and for enhanced accuracy of land classification in urban areas
Lidar (light detection and ranging)	—Provides detailed, bare-Earth elevation data and data on the height of vegetation and structures —Data collection unaffected by time of day or night, cloud cover	—U.S. and non-U.S. private contractors and firms	—Red River Valley, N.D./Minn. —North Carolina —Richland County, S.C.
Radar	—Data collection unaffected by clouds or time of day or night —Highly useful in studies of snow, ice	—U.S. government (NASA) —Non-U.S. remote sensing companies and government agencies —Private aerial remote sensing firms	

Remote Sensing Data," held in Boulder, Colorado, in January 2002. Organized by the Space Studies Board's Steering Committee on Space Applications and Commercialization, the workshop brought together representatives of federal agencies and the nonfederal public sector, technical remote sensing experts, university scientists and applications specialists, and representatives of the commercial sector for 2 days of panels, discussions, and presentations. The focus of the workshop was the ways in which remote sensing has been and can be used in state and local government, the role of the federal government in fostering public sector uses of remote sensing, barriers and bottlenecks to the expansion of remote sensing use and how to overcome them, and mechanisms to facilitate the adoption of remote sensing. The information exchanged at the workshop and the very lively discussions it engendered, together with the experience of the steering committee, form the basis of this report.

Many of the issues previously considered by the steering committee also proved relevant to the issues that concern state and local decision makers. In its first report, *Transforming Remote Sensing Data into Information and Applications*,[2] the steering committee examined institutional issues related to the adoption of remote sensing applications in new settings and to bridging the gap between raw remote sensing data and the information needed by decision makers. That report focused on remote sensing applications in the coastal zone, but the issues it examined have far wider relevance. Moreover, because many state and local governments have jurisdictional responsibility for some coastal issues, the coastal applications discussed in that report may be of interest to them (see Appendix A for the executive summary of the report).

The steering committee's second report, *Toward New Partnerships in Remote Sensing: Government, the Private Sector, and Earth Science Research*,[3] examined public–private partnerships for providing satellite remote sensing data for scientific research. Again, some of the issues discussed in the second report, though not directed at state and local government needs, can be instructive in thinking about remote sensing in the nonfederal public sector.

HISTORY

The technology of land remote sensing has advanced rapidly. Beginning in the 19th century with the photographic reproduction of landscapes obtained through a variety of devices that enabled overhead photography, the field expanded rapidly in the 1930s and 1940s with the development of airborne photog-

---

[2]Space Studies Board and Ocean Studies Board, NRC, *Transforming Remote Sensing Data into Information and Applications*, National Academy Press, Washington, D.C., 2001.

[3]Space Studies Board, NRC, *Toward New Partnerships in Remote Sensing: Government, the Private Sector, and Earth Science Research*, National Academy Press, Washington, D.C., 2002.

raphy. The commercial firms that obtained these photographic images were generally small and operated on a local rather than a national scale.

With the launch of the first civilian land remote sensing satellite in 1972 and the beginning of global-scale remote sensing, there was an expectation that state, county, and local governments would rapidly create applications for this new source of information, but for a number of reasons public sector applications were slow to develop. There were limitations on what could be accomplished with satellite data in urban areas because the resolution of the satellite imagery was lower than that of the airborne imagery upon which many cities had depended for so long. Few people who had technical experience with remote sensing data worked in the public sector, which in any case had established nontechnical means of obtaining the necessary information. In short, for many jurisdictions, there was no compelling reason to incur the added institutional and budgetary expense of introducing a new technology into their operations—a technology perceived as having limited practical applications.

There have since been changes in both the institutional and the technical capabilities of state and local governments. Management of spatial (geographical) data on terrains, land ownership, land use, and soils collected by state and local governments has become far more sophisticated in recent years. GIS software has become more flexible, powerful, and easy to use. Layers of digital data can be combined to form new information products. Many state and local governments now employ staff with geographic information science expertise and use GIS databases routinely.

Global positioning system (GPS) technologies also contribute to the geospatial resources available to the public sector. GIS and GPS technologies are particularly useful in urban applications and management. One of the advantages of these technologies is that they can easily be used in conjunction with remote sensing. Because remote sensing data can be georeferenced, they can be combined with topographic, land use, or tax data in a GIS database to provide information not previously available.

State and local governments have never before had such a broad array of land remote sensing data available to them. Table 1.1 lists some of the types of remote sensing data from which these governments can draw. The data can be obtained from an array of sources, such as the federal government, foreign governments, commercial satellite remote sensing companies, and aerial photography firms.

ORGANIZATION OF THE REPORT

Chapter 2 discusses how state, local, county, and regional governments are using remote sensing data; it examines cases in which remote sensing was used to solve real problems. Chapter 3 looks at how these governments addressed the problems that arose in connection with the introduction of remote sensing applications. Chapter 4 discusses the mutual benefits when federal and nonfederal agencies cooperate in remote sensing and examines the role of the private sector in meeting the remote sensing data and information needs of the nonfederal public sector. Chapter 5 contains the findings and recommendations of the steering committee.

Appendix A contains the Executive Summary of the steering committee's first report, which addresses barriers to developing remote sensing applications and proposes steps to address those obstacles.

2

Public Sector Applications of Remote Sensing Data

This chapter discusses some of the ways that state, local, and regional governments use remote sensing data and information. In a few cases, the use of satellite remote sensing data and information is becoming routine; in others, remote sensing applications are still experimental but are of growing interest to both technical staff and managers and decision makers in the public sector.

No one knows the information needs of state and local governments better than the people working in those governments. For this reason, the steering committee invited representatives of nonfederal jurisdictions to the workshop, where speakers discussed how remote sensing data and information are being introduced and integrated into their operations. Their accounts of the problems encountered and overcome in specific state and local governments can help other public officials better understand and envision ways that remote sensing might be of use in their own operations. In each case study, the steering committee examined how remote sensing data and information are used, the special advantages or characteristics of specific applications, and the problems encountered. The case studies are divided into three groups: local government applications, state government applications, and regional applications. Local government case studies are drawn from the experience of Baltimore, Maryland; Richland County, South Carolina; and Boulder County, Colorado. State case studies are from Missouri, North Carolina, and Washington. Regional studies are from Portland Metro in Oregon and the Red River Valley of North Dakota and Minnesota.

LOCAL GOVERNMENT USES OF REMOTE SENSING

Local government planning, management, and operations provide fertile ground for new applications of remote sensing data and information. Because they are responsible for geographically small areas that often have high population densities, city and county governments generally require high-resolution spatial data for a number of purposes such as cadastral or mapping applications, identification of changes in land use, and maintenance of the transportation infrastructure. Although local governments relied on aerial photography for high-resolution data in the past, some are finding that satellite remote sensing data are now available at similar levels of spatial resolution with broader spectral coverage. Many city and county governments already have in-house GIS capabilities; data from satellite remote sensing, like data from some forms of airborne remote sensing, can be used in conjunction with digital data available in existing GIS databases. The possibility of integrating remote sensing data into local GIS databases and using the databases in conjunction with locational GPS data has created opportunities for new types of information applications that were not possible using photographic remote sensing data alone.

Baltimore, Maryland: Introducing Remote Sensing to Urban Planning

The city of Baltimore initially looked to remote sensing to obtain urban data for a state map of forested areas in Maryland and to update the city's planimetric maps.[1] City planners had long depended on 19th- and early-20th-century maps for information on building and street locations in the city. In the 1980s and 1990s, as GIS software became easier to use, there were efforts in Baltimore to develop an integrated GIS, accessible to all agencies in Baltimore City, that would support planimetric map production and allow for GIS analysis. This effort was not initially successful. The early GIS products seemed crude next to the old-fashioned but artistically designed planimetric maps, and few people knew how to use GIS products for analysis. However, GIS offered the possibility of providing digital data for spatial analysis and decision making, as the static and inflexible planimetric maps, with their emphasis on fixed physical structures in the built environment, did not.

The city took its first steps into the world of remote sensing in order to obtain data for a map of vegetation cover. Maryland had acquired Landsat data to create a "greenprint" that identified the state's larger forests. The state's forest identification methodology, when combined with the resolution of the Landsat data, caused the urbanized areas of Baltimore City and Baltimore County to become blank spots on the state's forest map. Because Baltimore had forested areas in its

---

[1]A planimetric map is one where objects are mapped in their proper *x,y* geographic positions.

FIGURE 2.1 IKONOS 4-meter multispectral image of Patterson Park, Baltimore City, Md. Darkened background shows chlorophyll reflectance, which indicates areas of vegetation (e.g., grass and trees). SOURCE: Baltimore City Department of Planning.

parks, including small stands of original forest, city officials wanted urban forests to be mapped as well. With a grant from the Forest Service of the U.S. Department of Agriculture, Baltimore planning officials and the Maryland Department of Natural Resources (DNR) collaborated to obtain IKONOS imagery—high-resolution satellite images produced by the private sector that the city could use to map its forested areas (Figure 2.1).

Once the remote sensing data were in hand and DNR had extracted the vegetation from the image, Baltimore officials recognized that there were many other uses for the data, and today they are using them to create a land use map of the city for development, environmental, and social purposes and for updating the planimetric maps. The city is also planning for new applications in such areas as flood plain mapping, watershed planning, and identification of viewsheds.[2]

---

[2]A viewshed is a panoramic view from a specific point and is a function within GIS software that allows visualization of the data. A viewshed is the calculation of a line-of-sight map for a point location. Telecommunications groups use viewsheds to determine whether any ground interference might be present in the line of sight for cellular telephone towers.

The city of Baltimore, like many older urban areas, by law cannot annex land, so it is under pressure to use its existing resources more efficiently. Urban planners in Baltimore recognize that better spatial information on the physical and social attributes of the city, which can be provided by remote sensing data used in conjunction with other types of urban data in a GIS framework, can help in policy making for economic development.

Advantages Enjoyed by Baltimore

Baltimore has realized a number of advantages by introducing remote sensing data and moving into more advanced spatial data management for planning. The mayor is a firm supporter of data-based decision making and is interested in improving the city's spatial database and expanding the availability of maps and spatial data. This has been very helpful to urban planners trying to incorporate remote sensing data into decision making. Baltimore was also ultimately able to obtain support from the federal government for its initial acquisition of remote sensing data, so it did not have to make a cash outlay for a new type of data.

Issues Raised by the Baltimore Experience

Among the first issues Baltimore planners faced was how to finance the purchase of new remote sensing data and the construction of digital geospatial databases. The city had had no previous experience in purchasing this type of data. One major question was whether this should be a capital or an operating expense. Another was what agency or office should be responsible for the procurement. In the end, because of the difficulty in identifying funding sources, waste and wastewater funds in the city's Department of Public Works were used to create a GIS database that incorporated panchromatic imagery. Funding the city's data acquisition through a single agency raised a further problem, however—namely, the GIS was developed more for producing a planimetric map than for analysis and decision making. In part because of these issues, the Department of Public Works did not release the data to other potential users for 2 years.

Another problem was that Baltimore had limited technical experience in remote sensing and was unlikely to be able to increase its technical staff in the future. Many of Baltimore's employees with geospatial knowledge had learned about GIS and remote sensing technologies on the job. For some purposes, the city has issued contracts for technical services to groups at the University of Maryland or to individuals and consulting firms outside the city government. Though this is not uncommon in state and local government, it means that the city has to ensure that it has people on staff with enough training to define the city's data and information needs, manage external contracts effectively, and conduct quality control and assurance of the data.

Moreover, although the mayor of Baltimore supports mapping and the use of

GIS, the Baltimore experience suggests that it is also important to have a wide range of public sector managers and elected officials who are persuaded of the utility of digital spatial databases and remote sensing. These are the people who decide how to invest the city's funds and who serve as gatekeepers for the adoption of new technologies like remote sensing. It takes time and experience for local officials to understand and use new types of spatial data and information. Reaching and persuading these people—the real end users of the information—of the utility of remote sensing is critically important in implementing such applications in local government. At this stage, Baltimore urban planners believe that they need to demonstrate to government officials that multiple information products can be obtained from a single remote sensing image. They believe they can persuade decision makers of the utility of the data by providing them with cost-effective products that improve management and decision making in the city.

Baltimore planners attending the workshop also emphasized the need for continuity in remote sensing data: If the city has only one remote sensing data set (data valid at only one point in time) and cannot obtain comparable data in the future, the utility of the remote sensing data and information for policy and decision making will be limited. Continuity of data sources is critical for local government applications.

Richland County, South Carolina: Expanding the Uses of Remote Sensing

Richland County, which includes Columbia, the capital of South Carolina, is a leader in the use of remote sensing data and combines remote sensing with a wide array of other geospatial technologies. It finds remote sensing data, including satellite remote sensing, digital orthophotography, and lidar, critical in monitoring urban sprawl for smart growth policies and creating "fly-through" presentations that allow city and county officials to see the impact of specific construction and development projects (Figure 2.2). The county also uses remote sensing for managing its data on county infrastructure and monitoring the condition of pavements, floodplains, and land cover. It uses remote sensing for floodplain analysis; identifying, monitoring, and delineating wetlands; hydrologic modeling; and disaster response and mitigation. These many uses of remote sensing data in Richland County are integrated into the county's GIS system, and the resulting databases are enriched with digital photography and GPS.

Advantages Enjoyed by Richland County

Richland County has the people, the equipment, and the data in place to demonstrate the cost savings and superiority of remote sensing over other types of information in specific cases. For example, when a large telecommunications

FIGURE 2.2 Urban three-dimensional model developed by Richland County, S.C., using advanced remote sensing/GIS techniques. Surface elevations, vegetation dimensions, and building outlines and heights were calculated using raw lidar points and high-resolution orthophotography. Building facade textures were added using oblique digital photography.

firm was exploring the possibility of constructing a plant in Richland County, it asked the county to supply surveyors for a rapid assessment of land contours in the area—a 45- to 90-day job. Instead, using lidar data, the county GIS office was able to supply the necessary contours overnight, demonstrating cost savings of $140,000 for the county and technological efficiency to the potential investor.

Another advantage of Richland County's experience with remote sensing is that it assigned a county GIS office responsibility for the management of all geospatial data and information. This promotes the integration of multiple types of spatial data technologies and the creation of new data products for decision making. Equally important, it provides a visible point of contact for geospatial information within county government. The county GIS office is the institutional focus for budgeting, hiring technical personnel, and obtaining new data, equipment, software, and training in remote sensing and GIS.

Issues Raised by the Richland County Experience

If one of the strengths of the Richland County experience is having a single office to deal with all spatial data and technologies, including remote sensing, one of the issues that must be dealt with in the county is the division of labor among users of remote sensing data. County planners, public works engineers, economic development experts, and state officials who work with the county are all potential users of remote sensing data, but few of them have the background to understand when and where remote sensing can be applied in their own work. Without an understanding of the potential utility of remote sensing, they cannot know the potential benefits of the data for certain tasks or demonstrate them to others. Even remote sensing enthusiasts in local government admit there is a certain inertia in some sectors of their organizations such as permitting and land management. These sectors remain uninterested in remote sensing applications because traditional practices are effective and not easily replaced. In Richland County, the emphasis was on developing remote sensing applications for those areas where remote sensing could make an immediate contribution, not on serving all sectors of local government simultaneously.

For Richland County, as for many urban counties and cities, the valuable high-resolution data from commercial satellites are often too expensive for local governments to consider, particularly when the purchase of data is just one of the expenses associated with building a remote sensing and GIS capacity. Other expenses are for hardware, software, permanent staff, and training. In the past, local governments periodically spent large amounts for airborne data or surveying, which fitted into the experience of financial officers better than expenditures for a new technology that initially appears to be largely visual. The capacity to demonstrate financial and operational benefits—that is, savings of both money and time—was seen by workshop participants from Richland County and other local governments as essential to the successful introduction of remote sensing into local government. As state and local government revenues decline in the midst of a slow economic recovery on the national level, many workshop participants reported that public sector budgets are under strong pressures to constrain expenditures. As a result, cost savings will be a critical component of any new public sector remote sensing activities.

A related problem that arose in Richland County is how current commercial licensing practices can limit local government uses of commercial remote sensing data. Financial need leads many public sector agencies to seek ways of sharing the cost of new data across a number of agencies, so that the data are often used and reused for multiple purposes. This type of data sharing across agencies and government units can be limited by licensing restrictions if the data originate in the private sector, as do many of the high-resolution data sets for urban and suburban uses.

Another problem related to licensing came up in the discussion of the

Richland County experience—namely, that public agencies may be asked for their data by members of the public under the Freedom of Information Act (FOIA). Although disclosure is required by law, it is not clear whether the agencies must comply with such requests if the data are acquired under commercial remote sensing licensing restrictions.

Still other issues that arose in Richland County include the need for common standards across all counties so that the data in multiple jurisdictions can be compared and integrated into multijurisdictional databases; problems of data storage (Richland County already has one terabyte of data and is expanding its holdings); and legislation to limit the use of spatial technologies such as remote sensing and GIS in South Carolina.[3,4]

Boulder County, Colorado: Finding a Way

Boulder County began to use spatial data in 1987 to maintain and upgrade tax maps of parcels in the county.[5] From what was a modest beginning, the county now uses spatial data for a broad range of purposes, from locating prairie dog colonies to identifying wetlands to tracing fence lines. The county uses various types of remote sensing data in conjunction with GIS data in applications in public health, land use, parks and open space, road maintenance, and even redrawing precinct boundaries as population distribution changes.

Boulder County obtains its remote sensing data from Landsat; the Système pour l'Observation de la Terre (SPOT), a French remote sensing satellite; the Indian Remote Sensing Satellite (IRS); and aerial photos. The Indian remote sensing data are marketed in the United States by Space Imaging, Inc., a private firm, and SPOT data by another commercial entity, Spot Image. The county also creates its own data on roads, streams, and parcels.

---

[3]Examples from new legislation in South Carolina include:

- South Carolina Code of Laws, Title 40, Chapter 22, Section 40-22-20(23)(b) A photogrammetric surveyor determines the configuration or contour of Earth's surface or the position of fixed objects thereon by applying the principles of mathematics on remotely sensed data, such as photogrammetry.

- South Carolina Code of Laws, Title 40, Chapter 22, Section 40-22-225(E)(4) After June 30, 2004, no geodetic surveyor, photogrammetric surveyor, remote sensing surveyor, or GIS mapper may be licensed without meeting the requirements for education, length of experience, testing, or reciprocity criteria pursuant to this chapter.

[4]Fred Henstridge, "GIS & the Surveyor: Who Will Control the GIS?" *Professional Surveyor*, November 1999; Greg Pendleton, "Walking the Blurred Line of Geodesy," *GeoWorld*, June 2002; Fred Henstridge, "GIS for the Surveyor: GIS Opportunities for the Surveyor," *Professional Surveyor*, May/June 1997; Jerry McGray, "Geodetic Surveying Made Plain," *Point of Beginning*, January 2001.

[5]A parcel is a piece of land in any one ownership.

Advantages Enjoyed by Boulder County

In order to create a single county data interface for remote sensing and other spatial data applications in the public sector, Boulder County groups interested in spatial data organized BASIC (Boulder Area Spatial Information Cooperative). The cooperative has separate agreements with 30 to 40 public sector agencies, organizations, and jurisdictions in the region to provide unrestricted access to its data for BASIC member institutions. BASIC operates a distributed geospatial data and software capability for its members. Each participating agency or organization has its own GIS staff but draws on BASIC for software and some of its remote sensing data. When a public sector department or member jurisdiction does not have access to the common server or uses some form of dial-up access, BASIC may need to buy an extra copy of the data. In short, BASIC allows Boulder County to aggregate its members' needs for data (other than commercial, licensed data) into a single purchase, saving all the member jurisdictions money, avoiding duplication of data purchases, and providing a central location for data access and sharing. An added advantage of BASIC for data providers is that it creates a single interface for negotiations. Instead of having to negotiate data sales and agreements with 30 to 40 organizations in the region, data providers can deal with a single organization that is experienced and knowledgeable.

Another advantage of BASIC is that it provides a means of pooling resources—financial and human—among the members to reduce costs for all. When BASIC was faced with a price tag of over $150,000 for aerial remote sensing data, for instance, the organization reduced the cost by asking members to supply people to identify and mark section corners on the ground and to locate records of section corners held at the county courthouse before the aerial data were collected.

Issues Raised by the Boulder County Experience

For Boulder County, as for many other public entities, the cost of remote sensing imagery is a major issue. The problem is not only finding sufficient financial support but also persuading financial officers to approve the purchase. Boulder officials, like those in Baltimore, faced the problem of whether to categorize remote sensing data as a capital or an operating expense.

Licensing is another issue that complicates the use of remote sensing. Boulder officials told workshop participants that each type of remote sensing data came with different licensing restrictions. Differentiating among the various types of licensing restrictions has created confusion in local government agencies and among technical personnel. Localities like Boulder County may choose not to use licensed data if they cannot sell or share them with other local governments as a way to recover or mitigate data costs. The uncertainties surrounding what is allowable under the license, particularly when using multiple sources of data, some of which are licensed and some not, may constitute a disincentive to using the data.

STATE GOVERNMENT USES OF REMOTE SENSING

State governments also use remote sensing, although sometimes for purposes different from those of local governments. States are more apt to employ moderate-resolution Landsat data than high-resolution commercial data (whether from satellites or airborne sources) in their remote sensing applications, because they generally need data covering larger physical areas than cities and counties. Obtaining high-resolution data for large areas and preparing the data are too costly, and usually the phenomena in which states are interested can be observed with low- to moderate-resolution data.

Because state governments are also larger than local governments, with more employees and broader policy responsibilities, geospatial data activities can be isolated within agencies rather than shared across state agencies. As a result, there may be more duplication of effort from agency to agency or department to department in state than in local government. One state official told workshop participants that he had discovered that his state had purchased the same remote sensing image four times—each time by a different state agency unaware of the other purchases.

Finally, because sales taxes and income taxes are a major source of income in most states and local governments, their budgets are negatively affected by recessions and unemployment. State budgets, like local ones, are often subject to large changes from year to year, which may affect state capacity to purchase remote sensing data or launch new remote sensing programs.

Despite the differences between state and local governments in the use of remote sensing data, they have some patterns in common. For example, the critical need to persuade public sector managers of the importance and utility of remote sensing data and the value of being able to provide these data in a GIS framework were common to both state and local governments.

Missouri: The Value of Internal Data Coordination

The Missouri Department of Natural Resources (DNR) uses remote sensing data to monitor state wetlands. The state is now considering a broad array of other applications of this technology. New uses of remote sensing data being examined by Missouri include monitoring change in both urban and rural areas. In cities and on the urban fringe, the state is looking to remote sensing to identify development for taxation purposes. The Missouri Department of Economic Development estimates that the public sector loses approximately $10 million a year because it does not have information about when new homes are completed and when property taxes should be assessed. The use of remote sensing for detecting change is also being considered to trace the spread of oak blight in rural areas in the southern part of the state. Other possible uses of remote sensing are in watershed modeling, the siting of animal feeding operations, and understanding the impact of urban growth on agricultural areas.

Missouri also sees uses for remote sensing data in responding to natural hazards and unexpected events and estimating damages. The state faces potentially serious economic impacts from tornados and other extreme weather in both forested areas and urban settlements with extensive civil infrastructure. The nature of the damage can be estimated only if imagery obtained before the event is available to measure change after the event.

Advantages Enjoyed by Missouri

The state has a close relationship with state universities and often issues contracts for remote sensing data processing and analysis to groups in state universities. The flexibility of the contracts and the quality of the work at university research centers is a strong inducement to this practice, and it gives state officials access to advanced understandings and technologies in the field. However, relying on university scientists for significant remote sensing work under contract, instead of doing the work in state agencies, does mean that remote sensing activities are less visible to state managers and are therefore less secure.

Missouri, like Baltimore, also had the advantage of securing federal funds for its initial foray into remote sensing applications rather than having to use state funds for what would have been, in the context, an experimental methodology. The Missouri DNR has had a series of grants from the U.S. Environmental Protection Agency to delineate its wetlands. Initially, the state used Landsat data, although its more recent activities have been based on a combination of government and commercial remote sensing data, allowing state officials to compare the utility of each type of data for the wetlands task.

Issues Raised by the Missouri Experience

The steering committee was told that in Missouri, as in many other states, the initial uses of remote sensing were for new projects. The adoption of remote sensing data in Missouri differs in this way from the adoption of GIS, which was commonly employed throughout the state for numerous broad-scale uses. One of the implications of introducing a new technology into the public sector with project-specific applications is that other uses of the technology are often slow to emerge. As a result, in the case of remote sensing, there may be little coordination in the purchase or use of remote sensing data and technologies outside specific projects or agencies within the state. For example, when Missouri officials began to inventory the remote sensing data purchased by the state, they found that the same image had been purchased several times by different state agencies; to avoid a repetition of this situation, the state is now creating a regular process for cataloging its remote sensing data.

Missouri officials told workshop participants that cost, licensing, lack of understanding of remote sensing, and variable needs were major issues at the state level. They believe that flexibility in data sharing will be critical to the more widespread adoption of remote sensing in state agencies. The use of remote sensing data would be economically more attractive to public sector managers if state and local governments were able to coordinate data purchases and share remote sensing data across jurisdictions and agencies. Its cataloging initiative will make it easier for managers to locate data already purchased by the state.

The Missouri experience also emphasizes the value of having multiple data providers, federal and private, U.S. and foreign. Some state agencies are currently examining the use of SPOT remote sensing data because they perceive that Spot Image, the vendor, allows flexibility in public sector budgeting.

Missouri officials also emphasized the value of having state managers who understand and value remote sensing. Too often when remote sensing data are used in combination with GIS software, the remote sensing data lose their identity and are viewed simply as part of the GIS. State managers recognize the utility of GIS databases but often do not recognize that remote sensing data are part of the GIS. Similarly, when data processing and research are contracted out to state universities, public sector managers can lose sight of the importance of the activity to state operations.

Finally, Missouri officials suggest that states need more people trained in remote sensing. GIS training is now widespread, in part because training courses sponsored by GIS software firms are readily available. Remote sensing training is more difficult to obtain.

Washington: Remote Sensing for Federal Land Management

The Washington Department of Natural Resources (DNR) is beginning to use remote sensing to manage federal lands within the state. The DNR has responsibility for 5.6 million acres of trust land, 3.0 million of which were given by the federal government when Washington became a state. The land trusts include 2 million acres of timberland, about 1 million acres for agricultural grazing land, and about 2.6 million acres of coastal land on Puget Sound.

Unlike many agencies in state and local government that report to elected officials, in Washington the DNR is headed by an elected official, the Commissioner of Public Lands, who has responsibility for both managing trust lands and enforcing regulations throughout the state on lands that are in both public and private hands. Because the state receives income from sales of timber on trust lands, there is a growing emphasis on managing the trust lands using something closer to a business model. Stimulated by changes in budgets, requirements, and even management practices, the state is beginning to look to market signals for timing its timber sales, responding to demand at its peak rather than harvesting timber on an inflexible state schedule. This requires that the state have access to remote sensing data to know quickly where the needed timber can be found to respond to volatile market demands.

Washington is planning to use Landsat 7 data for mapping, change detection, and obtaining information for the state's fire protection program. It has commissioned development of a statewide mosaic of georeferenced, orthorectified Landsat 7 data from the EROS Data Center (EDC), a U.S. Geological Survey (USGS) facility in Sioux Falls, South Dakota. The dataset will be created from images obtained in 1999 and 2000. In addition, one county in Washington is using commercial 1-meter-resolution data—given to it by the data provider for demonstration purposes—to create a new vegetation map. The data are currently being used in a pocket GIS system planned for foresters in the field. If funds become available, the state is planning to explore the use of these data in a transportation corridor analysis pilot project to be done in collaboration with the State Department of Transportation.

Advantages Enjoyed by the State of Washington

One of the strengths of the Washington use of remote sensing data is the Washington State Geographic Information Council (WAGIC). This Landsat 7 consortium has been established by a number of state agencies, including the DNR, the Department of Transportation, the Department of Health, the Department of Fish and Wildlife, the Department of Assessor and Mapping, and the Department of Archeology and Historic Preservation, as well as several other state and county offices. The consortium creates a community of remote sensing data users in the state that draw from the same baseline data and pool their resources to cover the cost of the data. WAGIC is procuring a mosaic of remote sensing data to be used for multiple purposes throughout the entire state.

Another advantage is that Washington is able to draw on the most appropriate data for its work regardless of the data source. This means that it can use low-resolution Landsat data for statewide fire protection and high-resolution multispectral commercial data for projects within a single county. The budgetary stringencies that affect the state of Washington, like so many public sector remote sensing data users, are eased by institutional innovations and partnerships like WAGIC and by obtaining private data to demonstrate applications.

Issues Raised by the Washington Experience

Funding limitations that keep the state from taking full advantage of remote sensing technologies are a major problem in Washington. Planned pilot projects are being stripped to their essentials or postponed. The transition from manual black and white aerial photography to multispectral digital imagery can be difficult when there are funding limitations.

The state has experienced the common problem of a mismatch between the regulatory requirements and the data available to meet those requirements. The state elevation data obtained from USGS quad sheets can be as much as 50 years old and is of medium resolution (30 meters). GIS analysts often use these data in combination with high-resolution satellite imagery to create GIS products such as orthorectified products. However, merging older digital elevation models from the USGS quad sheets with high-resolution imagery can degrade the accuracy of the final data product, reducing the value of the remote sensing data.

North Carolina: Floodplain Mapping with a Purpose

North Carolina is subject to periodic hurricane damage that causes severe physical damage to the state and often has economic repercussions. Though many places on the East Coast suffered from Hurricane Floyd in 1999, in North Carolina Floyd was preceded by Hurricane Dennis and followed by Hurricane Irene, a combination that exacerbated the damage to property caused by Hurricane Floyd. The immediate damages in the state from Floyd were 51 deaths and a cost estimated to be $3.5 billion,[6] but the hurricane had longer-term impacts as well. For instance, many people moved out of North Carolina because of the flooding and the physical damage, a loss in population that created fiscal problems in later years.

In assessing the impact of the hurricane, it became clear that changes in terrain due to development and even flood damage in recent years meant that the state's flood insurance rate maps were badly out of date and that residents of North Carolina had often been using outdated and inaccurate information to insure their homes and property. In response to this situation, and to protect its residents and the state in future years, North Carolina decided to establish a flood mapping program to obtain, process, and disseminate accurate, current, and detailed data on elevation and flood hazards over the entire state and to provide these data in both digital and hard-copy formats. The state has committed itself to creating by 2006 digital terrain maps covering 48,700 square miles with a vertical accuracy of 20-25 centimeters.[7] Under its commitment to create the new digital flood insurance rate maps (DFIRM), North Carolina has contracted with two firms that will obtain lidar data using airborne sensors. Initial mapping is determined by river basins rather than counties because flooding is a function of

---

[6]John Dorman, "North Carolina Floodplain Mapping Program," presented at the NRC Steering Committee on Space Applications and Commercialization Workshop, "Facilitating Public Sector Uses of Remote Sensing Data," University of Colorado, Boulder, January 23, 2002; Brandon R. Smith, "Floodplain Fliers: North Carolina's Massive LIDAR Project," *Geospatial Solutions*, February 2002, pp. 28-29.

[7]Brandon R. Smith, "Floodplain Fliers: North Carolina's Massive LIDAR Project," *Geospatial Solutions*, February 2002, pp. 28-33.

rivers and river basins, not county boundaries. However, eventually the new maps will all be available by county. Once the initial data collection is finished, in 2006, the state plans to update the maps routinely thereafter.

The program is being done in partnership with the Federal Emergency Management Administration (FEMA), which has federal responsibility for flood insurance mapping. Under normal conditions, FEMA's budget would permit remapping only one of North Carolina's 100 counties each year; taking 100 years to map the entire state would create an untenable economic situation for a state that is subject to recurrent hurricane and flood damage. Under a new agreement, however, FEMA has assigned responsibility for flood mapping in North Carolina to the state, making it the first cooperating technical state under FEMA's Cooperating Technical Community partnership program. FEMA has agreed to match state funding of the mapping program.[8]

Advantages Enjoyed by North Carolina

North Carolina is benefiting in many ways from its remote-sensing-based flood mapping program. The DFIRM maps will be available to the public on the Internet, along with flood forecasts and inundation projections, in a format that is fully comparable across county boundaries. Earlier flood maps stopped at the county boundaries. A statewide GIS group, the Center for Geographic Information and Analysis, is obtaining memoranda of understanding from counties to establish a basis for sharing data across counties and municipalities. This will permit the state to incorporate existing spatial information in DFIRM maps as needed.

The mapping program will save money for the state. The USGS did a cost-benefit analysis of the likely impact of the DFIRM program in North Carolina and concluded that the state would gain $3.35 for each dollar spent on mapping and would lose about $57 million each year that went by without the maps.[9] These figures, together with the losses suffered as a result of Hurricane Floyd, helped state officials to recognize the financial costs of inaction and contributed to their willingness to move ahead with this program. But the North Carolina DFIRM maps will have other uses in the state as well. For example, flood maps can be overlaid with parcel tax maps, making it possible not only to model future flood impacts but also to use the DFIRM maps for community planning and development.

Another advantage of the North Carolina remote sensing mapping program is that it is based on a budgetary and institutional structure that anticipates mov-

---

[8]For additional background on the North Carolina Floodplain Mapping program, see <http://www.ncfloodmaps.com/>, accessed on June 28, 2002.

[9]John Dorman, 2002, "North Carolina Floodplain Mapping Program," footnote 6.

ing from a one-time to an operational mapping program. The one-time expense of the initial mapping is being covered with a combination of FEMA and state funds, but the state is also providing operating expenses for the program in future budgets, which will permit routine updating of the maps. Recognizing the need for a sizable one-time outlay and planning for ongoing operational expenses together ensure that the program will continue to deliver valuable information over time.

Issues Raised by the North Carolina Experience

DFIRM maps require significant digital storage space. North Carolina is purchasing 20 terabytes of data and will be using these in conjunction with data from other sources. There is a commitment to making the maps available to the public over the Internet through the North Carolina Floodplain Mapping Information System, but because of the storage constraints, some of the more detailed data will be archived at the USGS EROS Data Center in Sioux Falls, South Dakota, rather than being maintained online in North Carolina.

Since the maps are being developed by a state-federal partnership, another issue was whether to use meters or feet in the maps. (The federal government tends to use the metric system, and local governments tend to use the English system of feet and inches.) Still another issue to be addressed is differing practices among contractors related to the accuracy of the lidar-derived digital elevation information in terms of national mapping standards.

REGIONAL USES OF REMOTE SENSING

The number of issues related to population growth and the environment that must be discussed in a regional context is expanding. Moreover, natural disasters such as floods, hurricanes, and similar weather events do not affect discrete areas within county or municipal boundaries; they affect large regions that extend across and beyond political or administrative jurisdictions. Just as many governments cooperate in disaster response, they are also beginning to recognize the value of cooperating in the development of spatial data resources that can be used in disaster response and recovery across jurisdictions. Because of their capacity to provide comparable data over a large number of administrative units, remote sensing and GIS can be a valuable source of information for regional responses to growth and natural disasters, and for policy and decision making.

Portland Metro: Coordinating Land Use Data and Policies

The only elected regional government in the United States, Portland Metro is pioneering the use of remote sensing data in a regional spatial data framework in Oregon. Portland Metro, which encompasses 24 city governments, three counties, and several special-purpose management districts, is responsible for land use and transportation planning in the region. Because Portland set an urban growth boundary in 1979 that limits the physical expansion of the city, land use planning is a major policy focus in the region, and spatial data assume a critical role in policy discussions.

Metro maintains its spatial data in a regional GIS called the Regional Land Information System (RLIS), which contains information on land records and the built environment. Aerial photography has been used for about 10 years to inventory land use and natural resources within RLIS (Figure 2.3). This information has been supplemented with limited data on land cover from federal sources, principally FEMA and USGS. Satellite remote sensing data were first added to the RLIS in 1998, when Metro used Landsat imagery to obtain data on land cover in the region (Figure 2.4). The goal was to identify land that served as habitat for wildlife. Metro officials told workshop participants that their current goal is to

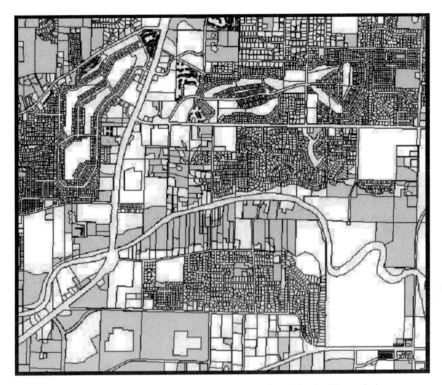

FIGURE 2.3 Annual vacant lots inventory of undeveloped land. The technique to produce the inventory uses visual interpretation of aerial photography to measure the land supply. SOURCE: Metro Data Resource Center.

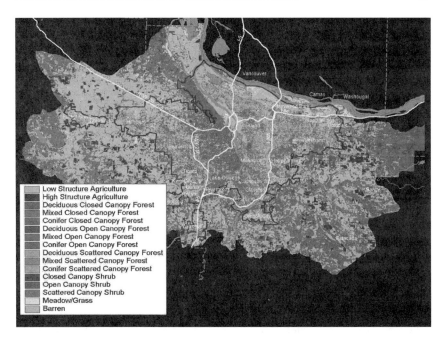

FIGURE 2.4 Land cover classification. This remote sensing analysis using Landsat data allows planners to model more complex relationships in the urban landscape. SOURCE: Metro Data Resource Center.

obtain higher-resolution data from commercial sources and to make these an operational part of the region's planning activities. At present, remote sensing data are used to inventory vacant lands within the urban growth boundary that can be used for industrial or residential purposes. They are also used to identify environmentally constrained lands, wetlands, and waterways.

Advantages Enjoyed by Portland Metro

One of the advantages enjoyed by Portland Metro is that the need for spatial data for urban growth planning led to the setting up of an operational budget for remote sensing data. The resolution of the data available to regional planners has improved each year, and Portland Metro has been able to assess the land supply more precisely because greater spatial detail is available from the higher resolution data.

Another advantage is that Portland Metro now has a common data infrastructure that is being maintained by the counties and municipalities that make up Metro. This collaboration has multiple advantages. It reduces the cost to local government of obtaining and maintaining data and information for decision making, and it means that the local governments within Metro that are working on land use policy can concern themselves with specific issues rather than arguing about data sources or quality.   One official said that in Portland Metro, data management is not a technical issue; it is about relationships.

Issues Raised by the Portland Metro Experience

The Portland Metro effort is complicated by the difficulties the local weather creates for obtaining remotely sensed images of the area.   Because of frequent rainfall in the area, it is often difficult to obtain cloud-free remote sensing images. Moreover, aerial photography is generally obtained in July and August, when the trees are in full leaf, making it difficult to use the images to observe aspects of the built environment like the transportation infrastructure or utilities.

Cost was identified as the most significant constraint to using satellite remote sensing in the Portland metropolitan area.   Metro officials would like to use high-resolution, multispectral commercial imagery because it gives significantly more information than aerial photography,[10] but the satellite data are considerably more expensive.   Metro officials hope that by pooling their resources and data needs across the region, it will be possible in the future to obtain the high-resolution data they need.

Red River Valley Flood: Remote Sensing and Disaster Response

In the spring of 1997, as the record snowfall of the winter of 1996-1997 began to melt and the waters ran north toward areas where the ice and snow were just beginning to melt, the Red River flooded its basin along the North Dakota-Minnesota boundary.   The floods were the worst on record, inundating farms and communities in the Red River Valley and forcing over 100,000 people to evacuate their homes.   The flood blocked regional transportation routes, disabled power systems, and disrupted water and sewage systems.   After the floodwaters receded, estimates of damages were between $1 billion and $2 billion.[11]

Standard USGS topographic maps are used to forecast floods by providing estimates of crests on the rivers of the Red River Valley.   Changes of only a few feet in the flat topography of the Great Plains can be significant in terms of flooding, but these changes cannot be discerned on USGS maps, which provide

---

[10]Multispectral data can be used to detect different types of vegetation and other ground targets, whereas aerial photographs cannot.

[11]Roger A. Pielke, Jr., "Who Decides?   Forecasts and Responsibilities in the 1997 Red River Flood," *Applied Behavioral Science Review*, 7(2): 83-101, 1997.

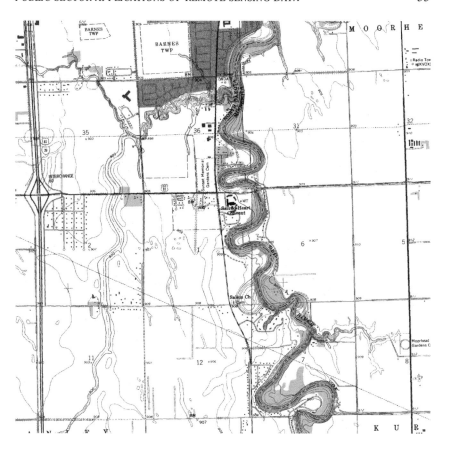

FIGURE 2.5   USGS contour map of Red River.   SOURCE: North Dakota State Water Commission.

only 5-foot or 10-foot contour lines (see Figure 2.5).   According to regional officials, difficulties in predicting the 1997 flooding demonstrated that maps with greater detail were needed to provide communities in the Red River Valley with better predictions of flood crests.   Regional officials considered several different technologies that could be used in developing new maps, including GPS survey-ing, aerial photography, and lidar.   In the end, two areas within the Red River Valley elected to create new contour maps using lidar—because of its speed, accuracy, and cost benefits—in combination with aerial photography.   Lidar was also selected because it can provide data for digital elevation models with 1-foot contour intervals, which can be used to reevaluate flood plains in the Red River Valley (see Figure 2.6).   Aerial photography was used to identify or verify ob-jects sensed by lidar.   For example, trucks, bales of hay, and tractors can be

FIGURE 2.6 Lidar digital elevation model shaded relief map of the Red River Valley. SOURCE: North Dakota State Water Commission.

identified in aerial photographs so that when they appear in the lidar data they can be recognized. In addition, aerial photography was used to help define land surfaces surrounding bodies of water, because lidar does not reflect off the surface of water.

Advantages Enjoyed by the Red River Valley

Although the mapping projects were led by the two North Dakota cities of Wahpeton and Fargo, each city organized a regional bistate partnership to obtain broader financial support for the work. Wahpeton worked with the North Dakota State Water Commission, the Corps of Engineers, FEMA, the Minnesota DNR, and the city of Breckenridge, Minnesota; the Wahpeton mapping project covered 65 square miles. Fargo created a partnership with the North Dakota State Water Commission; the Minnesota DNR; the Buffalo-Red River Watershed District;

Cass County, North Dakota; and Clay County, Minnesota. The Fargo project involved obtaining data for 138 square miles and processing the data for 42 square miles. The participation of multiple regional partners in both projects lowered the cost to each partner of obtaining the data.

Issues Raised by the Red River Valley Experience

One of the issues raised by these two remote sensing projects is the need for large data storage capabilities. Collecting remote sensing data creates very large digital files that require advanced software and hardware systems. Without adequate software and hardware, an investment in the creation of a large remote sensing database could be wasted.

Another issue raised by the Red River Valley experience is data ownership. A number of local and state government entities were involved in creating the new contour maps and remote sensing databases, and it was important to be able to distribute the data among the sponsors. The Wahpeton mapping project began first, and participants were surprised to learn that the lidar data obtained for the project were copyrighted by the contractor that had provided the data. When the contract for the Fargo project was signed, it specified that the data were to be owned by the purchaser. Regional officials told workshop participants that based on their experience, issues of data access and distribution should be addressed prior to signing a contract.

A third issue arising out of the Red River Valley mapping project is related to the ability of the lidar data to discern very small differences in elevation. If these differences are subject to even minor changes over time, such as the creation of contour lines caused by the furrows left by a tractor, the lidar data may have to be obtained annually. Local officials believe using remotely sensed data to keep the contour maps current will be a continuing responsibility.

3

Getting Started in Remote Sensing: Common Barriers and Bottlenecks

There was a remarkable convergence in the issues raised by the workshop presenters. In this chapter, the steering committee discusses the common problem areas identified by state, regional, and local government officials and looks at how public entities addressed them. Chapter 4 looks at how the federal government and the private sector facilitated the use of remote sensing data by state and local governments.

The workshop presentations emphasized that when public sector officials introduce remote sensing data and applications into existing or changing government operations, they must deal not only with significant technical issues but also with significant nontechnical issues. How these issues are addressed may determine whether a particular application of remote sensing in the public sector will succeed and whether remote sensing will be able to provide the information needed for management or decision making.

The issues fall naturally into several broad categories: (1) financial and budgetary constraints; (2) institutional, organizational, and political issues; (3) experience, skills, and training; (4) transitioning from photographic to digital data; and (5) licensing and data management.

A better understanding of the issues dealt with by governments that have already decided to adopt remote sensing may help transform the nonfederal public sector from what a workshop participant called one of the "largest untapped markets for remote sensing imagery" into a sophisticated, economically sustainable, technologically innovative sector for the use of remote sensing applications.

FINANCIAL AND BUDGETARY CONSTRAINTS

Financial constraints can have a major impact on state and local government operations. Workshop participants emphasized that fiscal constraints faced by jurisdictions in the nonfederal public sector are increasing. Many state and local governments are currently experiencing reduced revenues and even budgetary deficits. Because remote sensing is not often perceived as a must-have technology, the current fiscal situation in the nonfederal public sector could curtail or even eliminate any expansion of remote sensing activities in the sector.

The initial costs of developing an in-house remote sensing capacity can be considerable. There is, first, the cost of imagery itself. Though federal agencies provide data at a minimal cost and some types of data can be obtained free, the resolution of their data may not meet the needs of a public sector client that, particularly in urban and suburban areas, may require higher resolution. Such data must be purchased on the domestic or international market at commercial prices far in excess of the price of federal data.

In addition to data, however, there are other costs faced by state and local governments in developing an internal remote sensing capability, and these associated costs can add up to considerably more than the cost of the imagery. In the first report of this steering committee, *Transforming Remote Sensing Data into Information and Applications,* there was discussion of the many expenditures that must be assumed by a public entity intending to use remote sensing for the first time, and a recommendation that NASA should study the short- and long-term costs and benefits of developing remote sensing applications (see Appendix A). The steering committee reiterates the importance of this recommendation for informing state and local governments about the costs and benefits of using remote sensing data (Box 3.1).

These costs include, at a minimum, the initial cost and the cost of regularly updating hardware, software, and technical training for personnel. If trained personnel are not available on staff, introducing remote sensing may also include the cost of hiring new personnel with the appropriate technical and data skills or training existing GIS personnel. A viable alternative for many governments to establishing an in-house capability is to contract with a university or a private firm for remote sensing services. This approach may be expensive, however, and it does not build capacity within government for the horizontal diffusion of remote sensing applications.[1]

---

[1] The contribution of universities in providing education and training to the remote sensing workforce and in creating new types of applications is discussed at length in the steering committee's first report, Space Studies Board and Ocean Studies Board, National Research Council, *Transforming Remote Sensing Data into Information and Applications*, National Academy Press, Washington, D.C., 2001, pp. 42-44. The executive summary of that report is reprinted here as Appendix A.

BOX 3.1
Life-Cycle Costs of Remote Sensing Applications

In its first report, the steering committee stated as follows:

The full, life-cycle cost of developing and using remote sensing data products goes beyond obtaining the data and includes, among others, staff for data processing, interpretation, and integration; education and training; hardware and software upgrades; and sustained interactions between technical personnel and end users. . . .

Recommendation: NASA's Office of Earth Science, Applications Division, in consultation with other stakeholders . . . should mount a study to identify and analyze the full range of short- and long-term costs and benefits of developing remote sensing applications and the full costs of their implementation by public, nongovernmental, and other noncommercial users. In addition, NASA should support economic analyses to reduce the start-up costs of developing remote sensing applications.

SOURCE: Space Studies Board and Ocean Studies Board, National Research Council, *Transforming Remote Sensing Data into Information and Applications*, National Academy Press, Washington, D.C., 2001, pp. 3-4.

Capital or Operating Expense?

A related problem for a jurisdiction seeking to initiate or expand remote sensing activities is that it is not always clear how to budget for the purchase and use of remote sensing data. Remote sensing advocates in state and local government face problems in obtaining appropriate budgetary support for purchasing data. The Baltimore and Boulder County speakers at the workshop reported that they encountered uncertainty as to whether remote sensing data purchases should be considered a one-time capital expense or a routine operating cost. The problem could be exacerbated by the fact that the need for subsequent data for monitoring purposes or measuring change may not fit easily into the annual budget cycle of most public entities. Data may be needed annually or biannually or at some other interval that does not conform to or fit within the budget cycle. Public sector budgets are generally based on recurring costs; the purchase of a new type of data does not fit comfortably into a budget based on incremental deviations from the previous year's expenses.

Procurement

Finally, there is the issue of public sector procurement processes and their role in acquiring remote sensing data. Although public sector jurisdictions in the aggregate constitute a significant market for remote sensing data, the purchase of data is conducted by individual public sector units governed by complicated procurement regulations. There are 50 states and over 3,100 counties in the United States, all of which are potential customers with different data needs. There are many more towns and cities that might be interested in remote sensing data. Because the procurement of data for each of these units involves expenditure of public monies, it must be done according to processes that are transparent, fair, competitive, and based on principles that emphasize the lowest bid. Meeting such public sector contracting requirements can be a lengthy and costly process for both buyer and seller.

State and local governments have responded to these financial and budgetary problems in several ways. They have tried to justify the price of remote sensing data by comparing the costs of airborne remote sensing data with those of satellite data from the federal government and from the commercial sector and balancing cost estimates for specific tasks with anticipated uses of the data. Many public sector groups pool their data requirements and form cooperatives or centers that purchase remote sensing data for the use of all contributing member agencies or governments. This approach can reduce both the absolute cost and the transaction costs of obtaining data. This was the experience of cities in the Red River Valley, as well as Boulder County, Colorado, the state of Washington, and Portland Metro (see Chapter 2). Other governments, like that of Missouri, inventory remote sensing data that have already been purchased so that, if licensing arrangements permit, they can be used by the state for other purposes.

Procurement questions point to a more fundamental expectation about the use of remote sensing data in the nonfederal public sector. State and local governments may not anticipate new uses of the data beyond the specific project for which they were purchased. Though this one-time approach is encouraged by commercial licensing practices, one consequence is that public agencies do not then frame their data purchases in the broader context of multiple uses of the data.

A number of state and local governments, including Richland County, South Carolina, have reduced the costs of adopting remote sensing by building on the management capacity already in place for GIS and related geospatial data. By using existing staff and facilities, many jurisdictions can reduce the start-up costs of introducing remote sensing data. Not only do there appear to be economies of scale for the public sector entity that combines geospatial data activities into a single management unit, but there may also be benefits in being able to choose from among multiple types of geospatial data to select the most appropriate source of data for the information needed for management and decision making.

INSTITUTIONAL, ORGANIZATIONAL, AND POLITICAL ISSUES

Among the nontechnical constraints on the use of remote sensing data in state and local governments are (1) institutional issues related to the organization of remote sensing activities and (2) political and legal issues that can influence how the data are used. The workshop presentations repeatedly emphasized that state and local governments are subject to greater political pressures than the national government. It is not that party considerations govern public sector management. Rather, voters and elected officials are closer to operational decision making in states and localities than they are at the federal level. Elected officials are therefore more often able to influence expenditures on remote sensing, the uses to which the data are put, and the institutional structure within which remote sensing is managed.[2]

Use of Remote Sensing Data in the Courts

Problems with judicial acceptance of satellite remote sensing data as evidence in court cases can also influence the application and use of remote sensing in public sector entities. Because technological advances enable remote sensing satellites to identify objects smaller than a meter, the technology now permits detailed observation of personal movements, objects stored and used on private land, and even in some cases corporate production runs and inventories. This ability to observe has raised issues related to the right to privacy and protection from unreasonable searches. Perhaps because aerial remote sensing data have been available for a longer period of time, they have been upheld as evidence in legal cases, but space remote sensing data have not. Decisions about the use of satellite remote sensing data in court are expected to unfold as satellite imagery becomes more commonplace.[3]

---

[2]One opportunity for managers to influence the use of remote sensing data is in response to Government Accounting Standards Board (GASB) Statement 34, which established in 1999 new financial reporting requirements for state and local governments in the United States. One requirement is that nonfederal governments are to report on the physical condition and maintenance level of such infrastructure as roads, highways, bridges, and sewers. According to one county representative, remote sensing information could be a valuable tool for conducting inventory surveys to comply with GASB Statement 34. Compliance affects a city, state, or county government's bond rating and credit worthiness, an important factor for nonfederal governments that often finance infrastructure investments with municipal bonds. For information on the GASB and Statement 34, see <http://accounting.rutgers.edu/raw/gasb/repmodel/index.html>, accessed on August 6, 2002.

[3]For more on the use of aerial photography in court cases, see the discussion in a note issued by the NASA Inspector General's Office, "Remote Sensing and the Fourth Amendment: A New Law Enforcement Tool?" at <http://www.hq.nasa.gov/office/oig/hq/remote4.html> (accessed July 2002); Timothy J. Brennan and Molly K. Macauley, "Remote Sensing Satellites and Privacy: A Framework for Policy Assessment," Law, Computers, and Artificial Intelligence, 4(3):233-248 (1995).

Internal Advocates

The workshop presentations emphasized the importance of having an internal advocate for remote sensing when developing a successful public sector remote sensing program. This could be a person with program responsibility who needs remote sensing data to do the job, or it could be someone with programmatic responsibility for geospatial data and technologies. Regardless of that person's functional position in the government, his or her role as the internal advocate appears to be important both in advancing the use of remote sensing data and information and in fostering the building of capacity. Without such an advocate in-house, the initial budgetary, organizational, and technical barriers that must be overcome to establish the use of remote sensing data operationally can appear to outweigh the benefits. Moreover, the benefits themselves are usually not well understood. The internal advocate provides information to public sector managers and decision makers on what remote sensing can do and takes the initial steps to obtain data and the capacity to develop remote sensing applications.[4]

Organizational Location of Remote Sensing Expertise

Another institutional issue that must be addressed is where to place remote sensing expertise organizationally within the government. This issue affects the state or local government agency's ability to share remote sensing information with other agencies, and in turn affects the overall visibility of remote sensing applications. There appear from the presentations at the workshop to be several viable options, depending on the nature of the applications needed, the long-term

---

See also *Florida v. Riley*, 488 U.S. 445 (1988), in which the state of Florida convicted Michael Riley of possession of marijuana, using photographs taken from a helicopter hovering 400 feet above a greenhouse (with missing roof panels) on Mr. Riley's property. Mr. Riley appealed his conviction, arguing that the helicopter overflights, undertaken without warrant, violated his Fourth Amendment rights against unreasonable searches. The U.S. Supreme Court upheld his conviction, reversing both the Florida Court of Appeals and the Florida Supreme Court. Another important case supporting aerial searches is *California v. Ciraolo*, 476 U.S. 207 (1985) (affirming the right of police to observe visually a marijuana patch from a private plane within airspace available to the public).

On the use of digital information and data in court, see Jon L. Roberts, "Admissibility of Digital Image Data and Animations: Courtroom Concerns," *Advanced Imaging*, August 1995, pp. 101-102. Also see T. Slonecker, D.M. Shaw, and T.M. Lillesand, "Emerging Legal and Ethical Issues in Advanced Remote Sensing Technology," *Photogrammetric Engineering and Remote Sensing*, 64(6):589-595.

[4]Melanie Wallendorf, University of Arizona, "Adaptors and Adopters of Remote Sensing Data," presented at NRC Steering Committee on Space Applications and Commercialization Workshop "Facilitating Public Sector Uses of Remote Sensing Data," University of Colorado, Boulder, January 23-24, 2002.

strategies for using remote sensing, and existing government structures. It is necessary to address this issue even if a public sector entity decides not to conduct remote sensing work in-house but to contract it out to a local firm or university. In the latter case, there is a need for in-house remote sensing expertise for preparing, competing, and monitoring contracts. If an operational remote sensing capacity is set up within the public sector, remote sensing is likely to have a more visible and permanent role in both operations and policy discussion than when data are obtained through external contracts.

The presentations at the workshop suggested that state and local governments have responded to these institutional, organizational, and political issues in different ways. Most presentations emphasized the need to communicate regularly with elected officials, in addition to government managers and decision makers, about the benefits and uses of remote sensing data (a point emphasized also by the steering committee in its first report[5]). This is not only reasonable but also valuable in creating an interest in and support for remote sensing. It is even valuable to communicate with voters. Too often the contributions of this technology do not come to the attention of public sector leaders, or the role of remote sensing data is hidden or overlooked in substantive discussions of the information obtained from it. As one participant noted, no one ever lost an election because of remote sensing.

Some participants noted a tendency in speaking with elected officials and upper management to emphasize the pretty pictures produced by remote sensing at the expense of more utilitarian discussions of its applications. Although the images have an intrinsic appeal, these participants felt that support for remote sensing is stronger when elected officials understand the long-term informational benefits, including the economic benefits, of using remote sensing data and its role in the protection of life and property. Others felt that the images offered a way to attract the attention of elected officials and upper management but agreed that the story of remote sensing should not stop with the image. Highlighting the contribution of remote sensing by discussing successful applications was felt to be very useful.

---

[5]In its first report, the steering committee stated as follows:

> Because the utility of remote sensing data is in its information content, and the ultimate users of remote sensing applications are likely to be nontechnical decision makers who influence budget decisions, it is very important that the end users understand the potential, the advantages, and the limitations of remote sensing data.

See Space Studies Board and Ocean Studies Board, National Research Council, *Transforming Remote Sensing Data into Information and Applications*, National Academy Press, Washington, D.C., 2001, p. 43.

Experience, Skills, and Training

Discussions at the workshop emphasized that there are few formally trained remote sensing experts in state and local government. Many of the people who work in this field have moved into remote sensing responsibilities from technical positions that require GIS training and experience, as did the geographic information officer of Richland County, South Carolina. Others have moved into the field of remote sensing from analytical positions in offices that require the type of data that can be obtained only through satellite remote sensing, as in the state of Missouri and the city of Baltimore. Not surprisingly, although many remote sensing staff in the public sector are highly trained, some who have remote sensing responsibilities are self-taught, as was pointed out in the Baltimore case study.

However, when a public sector entity does not have enough technically trained staff, it can be at a disadvantage in drawing on available remote sensing resources. For example, at the planning meeting for the workshop, National Oceanic and Atmospheric Administration (NOAA) National Environmental Satellite, Data, and Information Service (NESDIS) officials reported that some state and local governments did not participate in a remote sensing grants program directed specifically at the nonfederal public sector because they lacked sufficient remote sensing staff.[6] In addition, the Baltimore and Portland Metro case studies illustrate the need for technically trained government workers who can manage remote sensing contracts that are outsourced to third parties.

To improve their remote sensing expertise, as was mentioned earlier, many state and local governments work closely with local university centers for remote sensing research and hire new staff from among their graduates. State and local officials at the workshop spoke of the advantage enjoyed by agencies that can draw on local universities with strong remote sensing programs, as is the case in South Carolina, Colorado, and Missouri.

Many workshop participants also mentioned a potential role for the private sector in remote sensing training and skills development. The role of the private sector in providing an extensive and widely available array of GIS training courses was seen as a viable and valuable model for the extension of remote sensing expertise.

In its earlier report, the steering committee recommended that remote sensing training be provided not only for technical personnel but also for managers and decision makers, noting that

---

[6]Since FY2000 the NESDIS Ocean Remote Sensing Program has conducted a public, peer-reviewed, competitive research and applications development grants program to increase the effectiveness of the use and delivery of remote sensing data. States and regional user organizations are encouraged to participate.

Whether in the public or the private sector, an organization's capacity to incorporate remote sensing applications into its operations depends on having either technical staff with the necessary skills and understanding to process the data and transform it into usable information, or knowledgeable staff who can manage contracts with external, value-adding service providers.[7]

Although the focus and purpose of the training would be different, it is useful for those who will be using information derived from remote sensing data to have some familiarity with the technology and some understanding of its potential and its limitations. Training or information sessions could be held at meetings that state and local managers and chief information officers are likely to attend, such as the National Association of Counties, the International City/County Managers Association, or the Environmental Consortium of the States.

TRANSITIONING TO DIGITAL DATA

The issues that must be addressed to improve and facilitate the adoption of remote sensing in state and local government are not all institutional. Some are technical. One of the most important of these is the need to develop digital mapping standards for remote sensing. State and local entities are governed by a number of statutes and regulations that dictate photogrammetric mapping standards, but data processing of digital remote sensing data is not yet regulated by any governmental standards. Algorithms are not standardized, nor are there standards for the acquisition of data. In part, this is because there are few standardized products or expectations for the products derived from remote sensing. In most cases, the products are designed for specific decision making or operational management activities in the jurisdiction obtaining the data. However, much of the growth in the application of remote sensing data in state and local government is likely to occur across jurisdictions. Moreover, throughout the federal government there is a growing emphasis on setting common standards for all types of digital data to facilitate the use of data from multiple sources for emergency management and homeland security.

Another issue raised at the workshop is that the federal government tends to work in the metric system and state and local governments use the traditional English system of feet and inches. This can complicate the use of remote sensing applications across federal and nonfederal agencies.

---

[7]Space Studies Board and Ocean Studies Board, National Research Council, *Transforming Remote Sensing Data into Information and Applications*, National Academy Press, Washington, D.C., 2001, p. 42. The report discusses the importance of educating nontechnical decision makers about remote sensing.

BOX 3.2
Traditional Mapping Standards Versus Digital Spatial Information Standards

The standards for specifying the parameters for constructing maps, which were developed as a result of the advent of aerial photography, are based on the characteristics of aerial cameras that use photographic film. The products have high spatial resolution, and the geometric correction of these images is complex because of the characteristics of the camera's optical system. As a result, the feature identification part of the prevailing standards is quite stringent, while the geometric standards allow for what is possible given the type of camera and the scale at which the data are acquired. The geometric correction of images acquired by digital sensors, on the other hand, is a relatively straightforward process compared with that of film images, but the resolution of digital imagery is limited by the size of an individual pixel when projected onto the ground. In general, a digital image that can meet the geometric specification of traditional mapping standards for a given scale is not able to meet the feature identification part of the standard for the same scale. The mapping standards based on digital imaging technology need to take into account the fact that there are fundamental differences between the characteristics of the spatial information carried in a digital image as compared with the corresponding film image.

Digital Standards

Requirements for geospatial information products are often stated in terms of the mapping standards developed for the production of maps from aerial photography rather than in terms of digital standards suitable for introduction into modern computer-based GIS. This makes it difficult for commercial remote sensing companies to respond to requests for proposals. Photogrammetric mapping standards are based on traditional aerial photography (analog) technology. Although these data are now often processed digitally, the mapping standards in common use for operational mapping are based on the use of photographic film to acquire the original data. Digital sensors, such as those carried by spacecraft, have fundamentally different characteristics (Box 3.2), which are difficult to reconcile with the traditional standards. This can significantly inhibit the use of remote sensing data in state and local governments, according to workshop participants.

LICENSING AND DATA MANAGEMENT

The licensing of remote sensing data and related issues of data access and continuity are critical for the economic and even the legal viability of using remote sensing data in state and local government. A number of participants in the workshop spoke of the need to amortize the cost of remote sensing data across multiple units, whether these be agencies within a state or locality, independent jurisdictions within a region, or partnerships of federal and nonfederal units. Although federal regulations regarding the use of remote sensing data, such as Landsat data, permit this type of data sharing, it is often prohibited by the licensing restrictions of private data providers. The barriers to sharing data that licensing restrictions present are one reason why many state and local governments continue to use aerial photography. Because the air photos are typically sold, not licensed, to the purchaser, they can be shared without penalty. Current data licensing agreements that prohibit relicensing and distribution are a disincentive to local governments interested in adopting products based on commercial satellite images.

Another problem raised by private sector licensing restrictions is legal. Many jurisdictions in the nonfederal public sector are subject to freedom of information requirements that they provide data to citizens requesting them. This problem was raised by several workshop participants, but there appeared to be no common response to it.

Data Sharing and Cooperatives

If the right to share data depends on licensing provisions, the capacity to share data is related to the technological skills, use of common standards, and organizational arrangements in state and local government. Remote sensing professionals in the public sector must have sufficient technical capacity to manage and provide access to data for multiple purposes and to combine them with other sources of spatial data such as GIS and GPS. Officials in Boulder County, Colorado, found that sharing data across jurisdictions is facilitated by having formal, written data-sharing agreements. These agreements are the basis for membership in BASIC. The state of Missouri found that having an inventory of remote sensing data purchased by state agencies and departments provided information about the status of state-owned data to individuals in other departments that might be interested in using the data. State and local universities also share nonproprietary data with local government agencies.

Data Continuity

Finally, data continuity is a critical issue for state and local governments. If a nonfederal public entity invests scarce resources in the purchase and use of remote sensing data, managers want the assurance that they will be able to obtain comparable data in the future. The investment in remote sensing is occasionally based on a one-time need for data, but more often it is premised on the belief that more data will become available for identifying and monitoring changes over time. For example, in the Red River Valley, even small changes in contours and elevation due to development and flood damage in earlier years can affect future flooding and water flow. New data will have to be obtained frequently, possibly annually.

If these data are obtained through a contract with an airborne remote sensing firm, the contract can be renewed each year and control over access to new data is in the hands of the state or local government agency. However, if the data for a remote sensing application are obtained from satellite sources, the issue of whether data will be available in the future depends on satellite availability and renewal. The decisions about these issues are made by federal agencies, private firms, and even international remote sensing data providers and are often based on long-term economic, policy, and strategic criteria rather than short-term needs for continuity in data for specific applications. Yet from the perspective of the current and future development of remote sensing applications and the development of a robust market for remote sensing data in state and local government, unanswered questions about future data availability and uncertainty about access to existing historic or heritage data can be a disincentive to investing in new applications.

4

Beyond State and Local Government

Although this report focuses on the use of remote sensing data and information in state and local government, the applications discussed at the workshop and the workshop planning meeting demonstrated that nonfederal governments are engaged in valuable collaborations with the federal government and the private sector. The interests and responsibilities of the federal government, the private sector, and the nonfederal public sector are clearly different, yet the case studies of state and local government use of remote sensing data presented at the workshop showed that all three sectors can benefit when they cooperate. Such cooperation makes it possible for state and local governments to introduce remote sensing data and information applications, which can in turn improve the quality of the data and information available for management and decision making at all levels of government. In previous chapters, the steering committee discusses collaboration within state and local governments; here it examines interactions of the nonfederal public sector with the federal government and the private sector.

The federal government is the most significant source of civil satellite remote sensing data used by state and local governments.[1] At present, there are

---

[1]William E. Stoney, "Summary of Land Imaging Satellites (with Better than 30 Meters Resolution) Planned to Be Operational by 2006," Mitretek Systems. Available online at <http://www.asprs.org/asprs/news>, accessed on November 22, 2002. This source lists U.S. and foreign land remote sensing satellites, both government and commercial. Information on U.S. meteorological satellites can be found at <http://www.noaa.gov/satellites.html>, accessed on November 22, 2002.

many remote sensing satellites in orbit (both land and meteorological) operated by the National Aeronautics and Space Administration (NASA) and NOAA. Meteorological satellites have long provided essential information on weather and climate to state and local governments, helping them anticipate extreme weather events, but the emphasis in this workshop was on land remote sensing. In this field, the Landsat series, which has provided a continuous flow of moderate-resolution land observation data for 30 years, is perhaps the satellite data source most commonly used in the nonfederal public sector. Data acquired from satellite sensors such as the Advanced Very High Resolution Radiometer (AVHRR) and the Moderate-Resolution Imaging Spectroradiometer (MODIS) are also increasingly used by state and local government. Some jurisdictions in the United States have also purchased images from France and India for use in remote sensing applications. Since the successful launch of two high-resolution land observation satellites, IKONOS and QuickBird, by private firms, state and local governments have also had the option of obtaining high-resolution data from the private sector in the United States. In some cases, instead of purchasing multiple commercial data and images for analysis over time or space, there have been attempts to use high-resolution commercial imagery to calibrate lower-resolution Landsat data for use in urban areas.[2]

THE FEDERAL GOVERNMENT AS A PARTNER

Its Role as a Data Provider

Because many regional and statewide applications of remote sensing for understanding land-based phenomena depend on Landsat data, workshop participants reported that the status of the projected Landsat Continuity Mission (the eighth Landsat satellite) and a guarantee of continuing public access to Landsat data are of great importance to officials in state and local government. The steering committee was told that continuity in the Landsat data stream is invaluable for monitoring and measuring change in public sector remote sensing applications. The cost of data is always a concern to local government, and the cost of Landsat data, although far lower than that of commercial data, is often too high for many jurisdictions, according to workshop participants, who said that an effective way to increase the use of satellite remote sensing data in state and local government would be to make Landsat data available at greatly reduced prices.

In addition to providing state and local governments with satellite remote sensing data, federal agencies also support the preservation and management of

---

[2]Christopher Small, "Multiresolution Analysis of Urban Reflectance," IEEE/ISPRS Conference on Remote Sensing of Urban Areas, Rome, Italy, paper 23, 2001. Available online at <www.LDEO.columbia.edu/~small/Urban/Urbref.html>.

TABLE 4.1 Major Federal Remote Sensing Data Centers and Archives

Data Center/Archive	Description
USGS National Satellite Land Remote Sensing Data Archive	Disseminates data from a 40-year archive of satellite remote sensing data of the land surface, including Landsat 7.
NASA Earth Observing System Data and Information System (EOSDIS)	Archives and provides active access to data from NASA Earth science missions and applications activities.
NOAA data centers (e.g., the National Climate Data Center, National Geophysical Data Center, National Oceanographic Data Center)	Archives long-term data, ensures quality and reliability of data, provides databases and data products for users, disseminates data.
NOAA National Coastal Data Development Center	Catalogs and makes publicly available coastal data from NOAA and non-NOAA sources, such as the data sets created by states (e.g., Texas, Maine, Mississippi).

extant and heritage remote sensing data through data archives and dissemination centers. These could be a valuable resource for state and local governments. The data centers most widely used by state and local governments are the National Satellite Land Remote Sensing Data Archive, operated by the USGS at EROS Data Center in South Dakota; NASA's Earth Observing System Data and Information System (EOSDIS); and NOAA's data centers, particularly the National Climate Data Center (NCDC) (see Table 4.1). NOAA is upgrading its data management infrastructure to accommodate a higher volume and rate of data production from the Terra, Aqua, and National Polar-orbiting Operational Environmental Satellite System (NPOESS) satellite systems. (To provide data for coastal applications, NOAA recently established the National Coastal Data Development Center, a centralized service to locate, access, and manage both NOAA and non-NOAA coastal data and to conduct hazard planning, fisheries management, and coral reef management.) Many of the federal government data centers make remote sensing data available at no charge or at the cost of filling the user's request, a policy that is attractive to local government officials, particularly when the data are to be used to meet federal government data requests. Unfortunately, because many of the data products available through these national data centers were originally intended for scientific and educational use, they often need additional processing before they can be used by state and local governments for management and decision making.

Federal Support for Remote Sensing Research and Development

The federal government makes a significant indirect contribution to state and local governments through its support for research and development. By strengthening the remote sensing teaching and research infrastructure in state universities through research grants and contracts, the federal government contributes to the improvement not only of the national science and technology base but also of the regional technical resources available to state and local governments. Many workshop participants said they work closely with their local universities, which are a source of consulting expertise and a locus for the training and education of new professionals in the field. For example, the workshop speaker from Missouri reported that the state issues remote sensing contracts to the University of Missouri at Columbia, Rolla, and Kansas City. In discussing the organization of remote sensing activities in Boulder County, the steering committee learned that the county has data sharing agreements with the University of Colorado Planning Department. Similarly, the state of Washington has established a test bed at the University of Washington that is being used to assess the effectiveness of P-band for penetrating vegetation.

Such federal agencies as NASA and NOAA have established partnerships with universities and other institutions in the study of remote sensing data applications. NASA's Regional Earth Science Applications Centers (RESACs) develop applications using remote sensing data to address regional problems. For example, the Mid-Atlantic RESAC at the University of Maryland works on regional land cover mapping, ecosystem modeling, and urban growth and planning. The Upper Midwest RESAC, a partnership of the University of Wisconsin at Madison, the University of Minnesota, Michigan State University, and others, works on remote sensing applications related to the use of agriculture, forestry, land use, and water data. NOAA's Coastal Services Center, in Charleston, S.C., works directly with state and local coastal resource managers and representatives from federal and nonprofit organizations to provide information, technology, and services that aid in managing coastal resources.

Although federal research support is based on scientific priorities, it also is concerned with linking basic research to applications. Emphasizing the link between basic research and applied data priorities could make federal scientific research more directly relevant to state and local government needs. As the steering committee recommended in its first report, "Resources, separate from funding for basic research, should be made available to federal agencies . . . for support of research on remote sensing applications and remote sensing applications derived from basic research. In addition, . . . agencies should establish joint research announcements aimed at fostering the development of applications for remote sensing data through basic research."[3] NASA's Office of Earth Science supports

---

[3]Space Studies Board and Ocean Studies Board, National Research Council, *Transforming Remote Sensing Data into Information and Applications,* National Academy Press, Washington, D.C., 2001, p. 5.

scientific research on remote sensing applications in a number of sectors, including community development and natural disasters and hazards, and fosters local government collaboration with university researchers.

Federal Grants to State and Local Government

The federal government plays a continuing role in fostering the use of remote sensing in state and local government. In a number of the case studies presented at the workshop, state and local officials said they had obtained federal grants that enabled them to acquire remote sensing data for applications. Baltimore, for example, used a grant from the U.S. Department of Agriculture's Forest Service to obtain IKONOS imagery to develop the "greenprint" of Baltimore. FEMA helped local government groups in the Red River Valley and the state of North Carolina obtain data for emergency management, and the Environmental Protection Agency helped Missouri obtain remote sensing data.

The Division of Applications in NASA's Office of Earth Science held four regional workshops in 2000-2001 to facilitate communication and collaboration on remote sensing among state, local, regional, and tribal users and commercial data providers and to provide guidance on data and technical support for remote sensing applications. These workshops also provided input for a Broad Agency [Research] Announcement (BAA) for funding projects that use NASA Earth science data and commercial remote sensing data in decision and policy making and other operational services.

NASA has also awarded grants to or arranged memoranda of understanding with professional societies that work with state and local government managers, such as the Western Governors Association, the Aerospace States Association, the National States Geographic Information Council, the Geologists Association, and the International City Managers Association, to expand the use of remote sensing data by the nonfederal public sector. NOAA representatives at the workshop planning meeting reported that the agency's NESDIS Ocean Remote Sensing Program initiated a competitive grants program in FY2000 that encourages state and regional organizations to seek support to improve the application of remote sensing data.

Federally sponsored workshops and reports are another vehicle for introducing remote sensing to state and local users. The regional workshops organized by NASA helped the agency understand who was using remote sensing data in state and local governments and for what purposes. NASA also supported preparation of a report by the National Conference of State Legislatures, *An Introduction to Geographic Information Technologies and Their Applications.*[4] Other studies

---

[4]Dena Sue Potestio, *An Introduction to Geographic Information Technologies and Their Applications,* Denver, Colo., National Conference of State Legislatures, March 2000.

have also addressed how state and local governments can use geospatial data and technologies.[5]

Other federal agencies provide support to state and local governments through the provision of data, research and development funds, the establishment of applications and technology transfer programs, and other services. NOAA officials at the workshop planning meeting described the agency's CoastWatch program, which links several states via regional offices into a national network for delivering environmental satellite data to end users. Similarly, the NOAA Coastal Services Center conducts studies on the use of and requirements for geospatial data for coastal resource managers, provides training for using the data, and offers a number of data products and services to its users.

Yet, despite the range of federal programs that offer state and local governments remote sensing data, services, or technical resources, many communities do not have the benefit of federal assistance in developing remote sensing data and encounter difficulties in working with the federal government.[6] There is a need for a continuing focus on technology transfer. Remote sensing technologies are changing rapidly, and their applications and user communities are expanding. Although there is an ongoing tradition of research on the efficacy of technology transfer practices and policies, this research has often focused on international development issues and is not always relevant to intergovernmental technology transfer or to such issues as the application of remote sensing and geospatial technologies in the nonfederal public sector. Moreover, research on technology transfer is rarely supported by the agencies that foster geospatial research or require data from state and local governments. Research on issues related to public sector user requirements and the implementation of remote sensing and related geospatial technologies in state and local government could foster the adoption of remote sensing.

Participants in the workshop emphasized that technology transfer to state and local governments would be facilitated by the development of common standards for digital data and that federal agencies could play a role in supporting community-wide efforts to develop such standards. (Concerns about the need to transition from mapping standards derived from aerial photography to digital standards were addressed in Chapter 3).

[5]Lisa Warnecke, Ronald V. Nanni, Zorica Nedovic-Budic, and William Stiteler IV, *Remote Sensing and Geographic Information Technology in the Nation's 50 State Forestry Organzations,* Syracuse, N.Y., GeoManagement Associates, Inc., 2002; Timothy Haithcoat, Lisa Warnecke, and Zorica Nedovic-Budic, "Geographic Information Technology in Local Government: Experience and Issues," *The Municipal Year Book 2001,* Washington, D.C., International City/County Management Association, 2001, pp. 47-57.

[6]Gail Elber, "A National Map or a Federal Map?" *Geospatial Solutions,* October 2002, pp. 18-20.

The Federal Government as a Consumer of Data

Because of its constitutional and legislative responsibilities for broad national and regional policy, the federal government imposes many types of data requirements on state and local government to help it meet these responsibilities. The 1990s saw an increasing number of data mandates issued by the federal government. Federal agencies required data from state and local governments for a wide array of projects, including mapping wetlands and floodplains and conducting inventories of vacant lands and natural resources. These data requirements can often be met most effectively through the use of remote sensing and other geospatial data and information technologies. However since federal data mandates are not often accompanied by appropriate funding, local governments tend to view them as an added staffing and budgetary burden.

Workshop participants gave examples of how federal data mandates can drive local governments to develop new data and information capabilities such as the mapping of urban forests in Baltimore. These new capabilities can in principle contribute to the overall development and advancement of technical skills in the public sector. According to workshop participants, however, competing demands on local government are so great and the budget limitations so severe that the next step, using newly acquired remote sensing technical skills for other public sector management and decision making purposes, is often not taken. Instead, the local government response to federal data mandates may be isolated in a single local agency, and neither the technical capabilities required to meet the federal requirements nor the data obtained for this purpose are transferred to other agencies of the local government.

Participants in the workshop said that federal data requirements are often passed down to lower levels of government without much consultation or even understanding of state and local capabilities. As a result, local government officials are forced to respond to requirements that are set without their input. Although they recognize that these requirements are the result of legitimate federal policy priorities and needs, local officials said that small alterations in federal information needs could, in many instances, make it easier for them to meet these needs with less budgetary stress. Local officials also told the steering committee at the workshop that it would be helpful if federal government agencies would provide technical training to meet data mandates. Economies of scale in training might permit federal agencies to do this at less cost than if each local government unit on its own were to devise or identify the training necessary to meet federal requirements.

The Federal Role in Emergency Response and Recovery

The experience of New York City in responding to the collapse of the World Trade Center on September 11, 2001, illustrates another important function of the federal government—the provision of geospatial information for emergency response and recovery efforts. Remote sensing data are an increasingly important tool for managing both natural and manmade disasters and emergencies. They have been used extensively in hurricanes and floods, but they can be used in other types of emergencies as well. Emergency management is an ongoing process that encompasses response, recovery, mitigation, and readiness. Predisaster imagery, if available, can be used to assess damage following a disaster. The U.S. Army Corps of Engineers and FEMA use several types of satellite data in emergency situations. They have found that the data are generally most useful at or below 10-meter resolution, although radar can be used to distinguish flooded from nonflooded land over areas that are broad and often difficult to reach. AVHRR and Landsat data, coupled with appropriate analytical techniques, can estimate the amount of moisture in snow and predict the amount of water that will be released when it melts.

In the World Trade Center tragedy, the effective use of geospatial data in the initial response and recovery effort was due to collaboration across the public, private, and university sectors. The availability of a city base map, prepared by a local city university in advance of the terrorist attack, was critical in providing New York City officials with the GIS framework in which to use remote sensing data obtained throughout the recovery effort. In assessing what made this data effort so successful, city officials report that it was the combination of existing geospatial data resources and new remote sensing resources that the federal government and others were able to provide after the event. It has been suggested that only the federal government can provide continuing service in high technology data and information resources for emergency response. Local governments will always be responsible for their own data for routine management, but emergency access to new forms of expensive high-technology data such as remote sensing data must be provided by the federal government or the private sector.[7]

Working More Effectively with the Federal Government

Despite the many types of interaction between federal agencies and state and local governments in obtaining and using remote sensing data and information, workshop participants said they frequently encountered roadblocks in working with federal agencies. At times these were caused by regulatory requirements; for example, federal agencies cannot become involved in disaster recovery or emergency management unless an area is formally declared a disaster by the President. At other times the problem is a lack of information and insufficient contact between the federal and the nonfederal public sector. Federal agencies

---

[7]Alan Leidner, OpenGIS Consortium, presentation, Columbia University, February 2002.

maintain remote sensing resources that could be of considerable benefit to state or local government agencies for both management and emergency purposes, but information about these resources can be difficult for state and local officials with little federal or remote sensing experience to obtain. From the workshop discussions, it appeared that state and local government officials lacked information on how to take advantage of federal resources. There is also a lack of communication. As mentioned earlier, state and local government officials said that they were rarely consulted about federal data requirements, so the requirements were more difficult to comply with than might have been the case had there been a consultative process when determining requirements. It appeared from the workshop discussions among both federal and state and local officials that greater communication and interaction could be helpful to both groups.

WORKING WITH THE PRIVATE SECTOR

Commercial firms, whether providers of satellite or airborne data or of value-added services, work closely with the public sector, providing data, information, or services for use in management and decision making. Yet workshop participants emphasized that the interactions between private sector firms and their public sector customers could be cumbersome and difficult. This is a situation in which the experience of small, locally oriented firms and large satellite data providers diverges. For example, as discussed in Chapter 3, small remote sensing firms have worked with local governments for years, but newly established national and international satellite data providers encounter problems when forced to negotiate small contracts with a multiplicity of local government units. Despite the potential size of the overall public sector market for data, the need to obtain many small contracts can be uneconomical for firms that must provide a return on investment.

The problems faced by private sector satellite image providers are related in part to the difficulties they experience in learning about opportunities to bid for work at the local government level. State and local government remote sensing is geographically and institutionally decentralized. There is no central information source that satellite firms can consult to find out which state and local governments are planning to use remote sensing data or are issuing RFPs. Decisions to work with local firms or universities are generally based on ease of access and the fact that it is good politics for the public sector to spend public monies locally. In addition, the local firm is likely to have better geographic and institutional knowledge of the region, which may be an asset in competing for a contract.

From the perspective of the public sector customer, the licensing requirements of satellite remote sensing firms can make it difficult for a public entity to use the data. Under some licensing requirements, data cannot be shared across agencies. In others, a government entity may be prohibited from using the data for multiple purposes (see, for example, Box 4.1).

BOX 4.1
Licensing Options for Commercial Remote Sensing Data

Space Imaging, Inc.

Licenses to use Space Imaging, Inc., remote sensing imagery can be obtained by

- A commercial business or government agency for a project for which the imagery is to be used;
- A commercial business at multiple locations or a limited number of related civil governmental agencies identified at the time of purchase for a project for which the imagery is to be used; and
- Federal civil agencies for use on a project for which the imagery is obtained.

Space Imaging, Inc., allows a licensee to

- Reformat the data product;
- Make one copy of the data product for internal archival or backup purposes;
- Distribute the data in a nonmanipulable format and on a noncommercial basis for research or publications purposes;
- Modify the product and make copies of the image product for internal use only;
- Distribute products derived from the Space Imaging, Inc., product; and
- Make products available to its consultants, agents, and subcontractors.

Space Imaging, Inc., offers 20 different license types allowing different combinations of sharing between federal civil agencies, DOD/Title 50 organizations, state and local governments, and international partners.

DigitalGlobe, Inc.

Licenses to use DigitalGlobe, Inc., remote sensing imagery can be obtained by

- A single organization that includes multiple users at multiple locations within a country (e.g., one corporation [excluding subsidiaries], one county government [all departments], one federal agency, one state or provincial government agency, or one city government [all departments]);
- Multiple organizations for multiple users solely within the corporations or government agencies within a single country identified at the time of ordering; and
- Schools and universities.

DigitalGlobe, Inc., allows a licensee to

- Make unlimited copies for internal use by the end user;
- Modify the data products to create derived works;
- Share the data with contractors for internal work;
- Release a limited number of hard copies for use on a noncommercial basis; and
- Publish on the Internet, with certain restrictions.

Another problem is related to public sector compliance with Freedom of Information Act (FOIA) requirements, which appears to be prohibited under commercial licensing restrictions. This puts state and local governments in an awkward and (possibly) legally vulnerable position. One response to these issues that was discussed at the workshop was for the public entity to provide the public or even other agencies with degraded data—that is, not to release the original licensed data but to supply a new data product that has been created from the licensed data.

A second approach is to seek a change in the formal licensing policy of the satellite image firms. At several points in the workshop discussions, it was suggested that the private sector is willing and able to make new licensing deals with potential public sector customers and that they have in fact already done so. Knowledge that this is possible could be extremely valuable for public sector geographic information and procurement managers. But it was clear at the workshop that knowledge of this possibility was not widespread, and the prevailing assumption that licensing provisions were immutable was a potential disincentive for some public entities to purchase commercial remote sensing data.

From the perspective of private data providers, the budgetary processes of state and local governments constitute a disincentive to doing business with those governments. Not only are public sector procurement processes arduous and time-consuming relative to the likely value of the contract but they are also dependent on public sector budget processes that are uncertain and subject to unexpected changes for political or economic reasons. One representative of a commercial remote sensing firm said his firm had learned that public sector budgets can suddenly be altered and that funding that appeared to be available can disappear.

Another problem is that the requirements for public sector remote sensing are often set in terms of photogrammetric mapping standards rather than digital standards. This reflects the fact that many public sector officials have long experience with photogrammetric images and do not know enough about digital data to state their data needs in terms of digital formats. It also reflects the fact that there are no widely accepted standards or sets of expectations for digital products. The algorithms are not standardized, and there are no standards for the acquisition of digital data (see "Transitioning to Digital Data," in Chapter 3).

The difficulties encountered by private firms seeking to work with state and local governments cannot be attributed solely to the lack of public sector experience or the complexity of working with public entities. Workshop participants raised the issue of whether the private sector had done enough to create a commercial market for high-resolution satellite remote sensing data in state and local governments. Some participants cited the role that GIS software manufacturers had long played in training individuals to use GIS and encouraging the growth of a market for GIS in state and local government as an example of how the private sector can contribute to the growth of an active user community. Even the representatives of private sector firms at the workshop recognized that they needed to do more to stimulate demand from the public sector applications community for remote sensing.

5

Findings and Recommendations

S tate, local, and regional governments have the potential to become a significant, and possibly the most sizable, market of civilian users of remote sensing data and applications. However, because they have responsibility for making a broad range of management decisions daily, officials in state and local government must have accurate, affordable, and accessible geospatial information that can be used alone or in combination with other types of statistical and administrative data and information. At its January 2002 workshop, the Space Studies Board's Steering Committee on Space Applications and Commercialization heard about wide-ranging and sophisticated applications of both satellite and airborne remote sensing data and information developed by state, local, and regional governments in the United States. Examples of remote sensing applications discussed at the workshop include the use of lidar for creating digital floodplain maps in North Carolina, the use of digital orthophotography and satellite remote sensing data to assess the impact of urban development in Richland County, South Carolina, and the use of Landsat data and high-resolution remote sensing data to map forested areas in Baltimore City and County. It also learned that the use of remote sensing data in management and decision making is currently uneven across the nonfederal public sector.

After several decades of little change in the sources of and technologies for using geospatial data, state and local governments have found that the field is now being transformed by technological change. Previously the mainstay of state and local government remote sensing applications, aerial remote sensing data, which were interpreted by photogrammetrists, are now used in combination with new types of data to meet the growing and increasingly more sophisticated needs of state and local governments. Many jurisdictions also now use digital geospatial data from a number of sources, including multi- or hyperspectral remote sensing data and data from radar, lidar, GIS, and global positioning systems.

The rapid pace of innovation in geospatial data technologies, and the corresponding increase in requirements for data management, in turn increase the technical, budgetary, and management demands made on state and local officials. Data collection, once routinely governed by contracts between a nonfederal public entity and its local airborne remote sensing contractors, is also changing. Some remotely sensed data are available only from global-scale data collection instruments, such as Earth observation satellites. These data are obtained from federal government, commercial, or non-U.S. satellite data providers, and their use is governed by a nonstandardized array of cost, licensing, and access restrictions. State, local, and regional governments may also have to employ technical consultants who can obtain and analyze remotely sensed data, including data from airborne sensors, some of which specialize in collecting bare-earth elevations that provide particularly good data on shorelines and bodies of water.

Critical elements in building the capacity of state and local government to use remote sensing include technical personnel, management and policy personnel, and hardware and software for data management and decision support. In an earlier report, *Transforming Remote Sensing Data into Information and Applications,* the steering committee suggested not only that technical personnel receive training in remote sensing applications, but also that managers and decision makers receive training in the ways the data can be used.[1] The steering committee reiterates the importance of this point for state, county, local, and regional governments, recognizing that this training may have to be tailored specifically to managers and decision makers and offered at meetings the managers would be likely to attend.

Although they provide many benefits, these recent technological changes also pose problems for the nonfederal public sector. State and local government responsibilities and expenditures are driven by budgets, laws, regulations, and politics and are often subject to practices and requirements that make it difficult to obtain and manage remote sensing data. For example, many local government officials report that their use of remote sensing data is now or will soon be limited by the capacity of their data storage facilities. Moreover, because state and local governments characteristically have stable staffing with little turnover, they may find it difficult to respond to technological change and related requirements for new expertise for implementing remote sensing applications. Increasingly, state and local governments are facing tight budgets and even shortfalls, which makes

---

[1] Space Studies Board and Ocean Studies Board, National Research Council, *Transforming Remote Sensing Data into Information and Applications,* National Academy Press, Washington, D.C., 2001, p. 43.

it difficult for these jurisdictions to afford the array of hardware, software, and personnel necessary to adopt new technologies for using geospatial data and information.

The steering committee found that the adoption of remote sensing data and applications was often related to having a strong advocate for the new technology in local government. This person could be an elected official, as in Baltimore, where the mayor is very supportive of improving the city's maps. More often the advocate is a government employee with a technical understanding of the potential value of the data for management and decision making, a person enthusiastic about the potential of remote sensing to contribute to the nonfederal public sector and willing to work with officials at all levels in the government.

The decision to invest in remote sensing data requires acceptance at multiple levels of state, local, or regional government. As one workshop participant noted, no politician ever lost an election because of remote sensing. But decisions about adopting and continuing to use remote sensing inevitably involve both technical and operational managers (including budget managers) and ultimately must also reflect the priorities of elected officials. Among the workshop's case studies of the use of remote sensing by state, local, and regional governments, two such uses—North Carolina's commitment to update and modernize its floodplain maps, and the formation of multijurisdictional consortia in the Red River Valley of the Upper Midwest to create new contour maps of the region—were the direct result of natural disasters. The prospect of recurrent losses in the range of billions of dollars undoubtedly influenced the policy decision to spend significant funds on new remote sensing data and applications that would improve regional response-and-recovery efforts in case of future hazards.

Persuading nontechnical public sector managers and elected officials of the value of remote sensing data and information is dependent on producing cost-effective information products. Although several participants in the workshop warned against reducing public presentations on remote sensing to "pretty pictures," the steering committee found that the images themselves have value in attracting the interest of nontechnical managers and officials. Once their interest is gained, officials are more likely to consider the utility of the data. Examples of the financial and management benefits of remote sensing data are also very persuasive in making a case for the use of remote sensing information.

Drawing on the experiences of state, local, and regional governments already using new remote sensing data and related geospatial technologies, the steering committee found that increased use of remote sensing data and applications in the nonfederal public sector is also related to (1) improvements in the management and efficient use of geospatial data, (2) creation of an effective nonfederal public sector market for remote sensing, and (3) cooperation between federal and nonfederal government agencies and entities.

IMPROVING MANAGEMENT AND EFFICIENCY

The state, local, and regional governments that have successfully begun to use remote sensing data and information for management and decision making have often been forced to reexamine certain types of management practices so as to reduce the cost and increase the efficiencies of developing remote sensing applications. It is administratively and economically advantageous for state and local governments that are considering the use of new remote sensing technologies or are just beginning to use remote sensing data and applications to learn from the organizational practices of governments that have already demonstrated successful adoption of remote sensing applications.

Geospatial Data Management

Based on its analysis of the case studies presented at the workshop, the steering committee found that effective management of geospatial data can contribute significantly to state, local, and regional government adoption and use of remote sensing data. Some state and local governments reported at the workshop that purchases and users of remote sensing data are spread through several departments. In some jurisdictions, there were multiple purchases of the same data and ineffective management and utilization of those geospatial data resources.

Because of the increased convergence of digital geospatial data such as satellite and airborne remote sensing, GIS, and even global positioning systems, jurisdictions can manage their use of geospatial data more efficiently and with less redundancy under a single administrative entity rather than separately in different departments or agencies within the jurisdiction. Whether administrative responsibility rests with a GIS coordinator, a chief information officer, or a geospatial data manager is less important than that the data, technologies, and development of applications are managerially linked. Such centralization will also facilitate the adoption of national and international spatial data standards.

Recommendation 1: A state, local, or regional government should consider making a single unit responsible for managing its geospatial data, information, and technologies.

Cross-Jurisdictional Remote Sensing Data Cooperatives

The cost of obtaining and managing remote sensing data can be prohibitive for state, local, and regional government departments or agencies, particularly during a period of budget shortfalls. The steering committee found that some governments in the nonfederal public sector have successfully joined to form local or regional cooperatives or consortia that purchase remote sensing data for all members of the group. This arrangement allows them to pool their resources and to amortize the cost of obtaining and storing data across multiple jurisdictions, providing efficiencies in monitoring contracts, licensing, and managing the data. In short, remote sensing or geospatial data cooperatives can improve data utilization and sharing and can minimize taxpayer expense.

Data cooperatives can also help small jurisdictions to manage remote sensing and other digital data by providing data storage capacity and fostering the creation of common datasets and GIS base maps for regional or cross-jurisdictional purposes. The existence of common base maps permits both governments and their constituents to focus on the issues at hand rather than on data quality and differences.

Recommendation 2: Public officials responsible for obtaining and using geospatial data should examine the benefits of forming multijurisdictional consortia or cooperatives to reduce duplication of cost and effort.

Procurement Processes

The purchase of remote sensing data does not fit easily into state or local government procurement processes. Governments in the nonfederal public sector often have little expertise in purchasing remote sensing data, particularly satellite remote sensing data. In addition, the procurement process can be lengthy and time-consuming from the perspectives of both the data vendor and the public sector official who will be using the data. It can also be derailed unexpectedly because of political or budget changes that are unrelated to the anticipated use of the data.

In natural disasters or emergencies, the time required for normal procurement processes can make the timely purchase of remote sensing data impossible. The experience of New York City offers an alternative approach. By establishing a long-term purchase agreement with a local university, New York City had the flexibility it needed to obtain geospatial data in response to the events of September 11.

A separate procurement problem faced in working with federal government agencies is that differences in the fiscal year between the federal and most nonfederal public sector entities create additional problems.

From the perspective of budget officials in state, local, and regional governments, remote sensing data are difficult to categorize for accounting purposes. It is not always clear whether the cost of the data should be seen as a capital or an operating expense. Moreover, expenditures for remote sensing data are likely to occur unevenly within and across fiscal years—a source of problems, given that budgets for state and local jurisdictions generally assume marginal changes in recurring costs, and state and local governments cannot carry over expenditures from one budget year to the next.

Recommendation 3: State and local government budget and procurement practices should be examined and modified, if necessary, to facilitate acquisition of multiyear remote sensing data.

An independent body such as the Government Accounting Standards Board—a private, nonprofit institution that develops reporting standards for state, local, county, and other nonfederal government entities—or another independent accounting organization could be consulted for input on how to account more effectively for expenditures on remote sensing data.

Recommendation 4: State and local governments should explore the feasibility of establishing long-term purchase agreements with local institutions or vendors to give themselves flexibility in obtaining remote sensing data.

CREATING A MORE EFFECTIVE PUBLIC SECTOR MARKET FOR REMOTE SENSING DATA

A large and active public sector market for remote sensing data and information will provide economies of scale for governments seeking cost-effective remote sensing applications and for those in the public and private sectors who supply data and services to state and local governments. The steering committee learned that there are many ways that a more active and effective market for state and local applications of remote sensing data and information can be created, including by developing standards for digital spatial data and information products, encouraging the private remote sensing industry to build a market in the nonfederal public sector, and establishing opportunities to advertise state and local remote sensing data requirements to remote sensing data and service providers.

Standards for Digital Spatial Data and Information Products

Many state and local governments are adept at dealing with photogrammetric standards for spatial (mapping) data. However, the increasing dominance of digital data means that common standards are needed for digital spatial data and information as well as remote sensing data products, which are inherently digital. The advantages of having commonly accepted digital spatial data standards are lower costs, improved ability to use the data for multiple purposes, standardization of technical training, and quality assurance. The adoption of digital data standards would require that procurement regulations be revised for many state and local entities. A coordinating body composed of key stakeholders and led by a federal agency could develop digital spatial data standards. Those federal agencies involved in this effort could identify which agency should take the lead.

Recommendation 5: The U.S. government, in collaboration with professional organizations, state and local governments, and vendors, should take the lead in establishing standards for digital spatial data and information products.

Private Actions to Build a Public Sector Market

Although commercial remote sensing image providers recognize the potential economic significance of the nonfederal public sector market for remote sensing data, they often do not do enough to stimulate the development and growth of this market. Many workshop participants pointed to the successful efforts of GIS software companies to increase the numbers of trained technical personnel and thus increase the demand for their products through training programs and national conferences.

Recommendation 6: To help remedy the lack of trained remote sensing personnel in state and local governments and to raise awareness of the advantages of working with satellite remote sensing data, commercial satellite data providers and remote sensing digital image processing vendors should look to GIS software companies as models for building the state and local government market.

Licensing

The licensing provisions of commercial satellite data companies seem restrictive, offering little flexibility to state and local governments. Commercial licensing provisions, if strictly followed, can add to the cost of purchasing remote sensing data. In some cases, the licenses may lead to redundant purchases of the same data within a single jurisdiction.

Licensing restrictions also affect the ability of state and local governments to resell geospatial data, a practice that many agencies use to leverage the cost of developing data on roads, parcels, or other local features and the cost of purchasing aerial photography, for instance. In these situations, strict licensing provisions for state and local governments constitute a disincentive for jurisdictions to purchase data from the private sector. At the workshop, representatives of private sector satellite firms suggested that it is possible to negotiate new licensing agreements based on individual state and municipal needs. However, officials from state and local governments who use remote sensing data indicated at the workshop that they were not aware of this potential flexibility.

Recommendation 7: Private sector providers of remote sensing images should offer standard information about flexibility in their pricing policies, ensuring that the information is widely available, especially information about jurisdiction-wide site licenses or long-term purchase agreements for state and local governments.

Opportunities to Work with the Public Sector

The steering committee found that there is no clearinghouse or single source of information on the remote sensing data requirements of state, local, and regional governments. The absence of a central information source on public sector data requirements restricts the market for services and data to firms in the immediate area or to those that have personal contacts with the public sector entity that is contracting for remote sensing data or services. If all the sources of data were local, such a practice might be reasonable, but remote sensing data are increasingly produced nationally or internationally, and involvement of the larger potential contracting community may help to keep pricing competitive.

Recommendation 8: Associations of state and local governments should establish national or statewide opportunities/forums for state, local, and regional governments to advertise their needs for remote sensing data.

COOPERATION BETWEEN THE FEDERAL AND NONFEDERAL PUBLIC SECTORS

The steering committee found widespread cooperation between federal agencies and state, local, and regional governments in developing remote sensing data applications. However, although this cooperation assisted both agencies and governments in doing their jobs better than they could have done alone, it focused on remote sensing applications that were developed for specific government programs rather than on general support to state and local governments that seek help in obtaining and using federal remote sensing data. There was little evidence of systematic attempts by federal agencies to provide a point of contact and means of facilitating collaboration with nonfederal governments in the development of remote sensing applications.

Recommendation 9: Federal agencies should have a formal point of contact for representatives of state and local governments that need technical assistance or want to identify sources of financial assistance for their use of remote sensing applications.

Appendixes

A

Transforming Remote Sensing Data into Information and Applications
Executive Summary

O ver the past decade renewed interest in practical applications of Earth observations from space has coincided with and been fueled by significant improvements in the availability of remote sensing data and in their spectral and spatial resolution. In addition, advances in complementary spatial data technologies such as geographic information systems and the Global Positioning System have permitted more varied uses of the data. During the same period, the institutions that produce remote sensing data have also become more diversified. In the United States, satellite remote sensing was until recently dominated largely by federal agencies and their private sector contractors. However, private firms are increasingly playing a more prominent role, even a leadership role, in providing satellite remote sensing data, through either public-private partnerships or the establishment of commercial entities that serve both government and private sector Earth observation needs. In addition, a large number of private sector value-adding firms have been established to work with end users of the data.

These changes, some technological, some institutional, and some financial, have implications for new and continuing uses of remote sensing data. To gather data for exploring the importance of these changes and their significance for a variety of issues related to the use of remote sensing data, the Space Studies Board initiated a series of three workshops. The first, "Moving Remote

NOTE: The executive summary reprinted in this appendix is excerpted from Space Studies Board and Ocean Studies Board, National Research Council, *Transforming Remote Sensing Data into Information and Applications*, National Academy Press, Washington, D.C., 2001.

Sensing from Research to Applications: Case Studies of the Knowledge Transfer Process," was held in May 2000. This report draws on data and information obtained in the workshop planning meeting with agency sponsors, information presented by workshop speakers and in splinter group discussions, and the expertise and viewpoints of the authoring Steering Committee on Space Applications and Commercialization. The recommendations are the consensus of the steering committee and not necessarily of the workshop participants.

Rather than trying to cover the full spectrum of remote sensing applications, the steering committee focused on civilian remote sensing applications in the coastal environment.[1] The workshop featured three case studies in coastal management involving (1) the application of Sea-viewing Wide-Field-of-view Sensor (SeaWiFS) data in monitoring harmful algal blooms, (2) the use of airborne lidar bathymetry for monitoring navigation channels, and (3) the use of both satellite and aerial remote sensing to identify sewage outflows. All three provided detailed information on the applications as well as problems encountered in developing them, allowing the steering committee to learn from the real-world experiences of particular users.

In addition, participants in five workshop splinter sessions—on education and training, institutional, technical, and policy issues in technology transfer, and user awareness and needs—identified and discussed more general barriers and bottlenecks that interfere with the development of remote sensing applications and also explored ways to overcome such problems. Plenary presentations focused on research on technology transfer; science and policy issues in the coastal zone; a comparison of remote sensing technology transfer with respect to geographic information systems and the Global Positioning System; and new directions in the use of remote sensing data. This material provided a basis for much of the steering committee's analysis and figured significantly in its development of the report's findings and recommendations.

BASIC OBSERVATIONS

To encourage finding more effective ways to develop new and useful applications of remote sensing data, the steering committee considered barriers to as well as opportunities for developing successful applications through the transfer of knowledge and technology.[2] Its examination of the remote sensing technology transfer process led to the identification of a number of gaps that must be bridged in order to develop effective civilian applications:

---

[1] Although a great deal of excellent work on operational applications has been done within the defense community, those developments were independent of civil remote sensing in terms of both budgets and technologies and hence they are not within the purview of this report.

[2] The steering committee approached technology and knowledge transfer in terms of the application of remote sensing data and images in the public, private, and not-for-profit

- The gap between the raw remote sensing data collected and the information needed by applications users. Users need information, and the process of transforming data into information is a critical step in the development of successful remote sensing applications.
- The gap in communication and understanding between those with technical experience and training and the potential new end users of the technology. Producers and technical processors of remote sensing data must be able to understand the needs, cultural context, and organizational environments of end users. Education and training can also help to ensure that new end users have a better understanding of the potential utility of the technology.
- The financial gap between the acquisition of remote sensing data and the development of a usable application. The purchase of data is only the first of a large number of steps affecting the cost of a successful application. An organization, commercial firm, or government agency that wants to incorporate remote sensing applications into its operations must be prepared for a long-term financial investment in staff, ongoing training (both technical and user training), hardware, and software, at a minimum. Alternatively, the potential user organization should be prepared to purchase these services from a value-adding provider.

Another recurring theme in workshop discussions was the need for data continuity. In light of the heavy, up-front investment required to develop and use remote sensing applications, organizations as well as individual users have to be assured of a reliable and continuous source of both data and information.

FINDINGS AND RECOMMENDATIONS

Life-Cycle Costs

Finding. The full, life-cycle cost of developing and using remote sensing data products goes beyond obtaining the data and includes, among others, staff for data processing, interpretation, and integration; education and training; hardware and software upgrades; and sustained interactions between technical personnel and end users (see Chapter 3). Although many of these costs are incurred at the time a technology is first employed, the life-cycle costs and benefits of remote sensing applications are not well understood.

Recommendation 1. NASA's Office of Earth Science, Applications Division, in consultation with other stakeholders (e.g., agencies that use remote sensing data, such as the U.S. Geological Survey, Department of Transportation, Envi-

---

sectors (regardless of whether they were produced by public or private sector image providers). These applications may depend on data from either the public or the private sector. Similarly, the process of technology transfer can take place within or across government agencies, between the public and the private sectors, within the private sector, and between the private or government sectors and the not-for-profit sector. At issue is not where the data originate or who uses them, but rather how to develop useful, operational applications.

ronmental Protection Agency, and U.S. Department of Agriculture; private companies; state and local government users; and not-for-profit institutions), should mount a study to identify and analyze the full range of short- and long-term costs and benefits of developing remote sensing applications and the full costs of their implementation by public, nongovernmental, and other noncommercial users. In addition, NASA should support economic analyses to reduce the start-up costs of developing new remote sensing applications.

Education and Training

Finding. Training is an integral component of efforts to bridge the gap between remote sensing professionals and end users (see Chapters 3 and 4). Remote sensing involves sophisticated technology, and specialized training is required to process the data, convert it into information, and interpret the results. Many agencies and organizations either lack the financial resources to provide such training or do not understand the importance of periodic retraining for technical staff.

Recommendation 2. Federal agencies such as NASA, the National Oceanic and Atmospheric Administration (NOAA), the U.S. Department of Agriculture, the U.S. Geological Survey (USGS), and others should provide the seed funding for developing remote sensing training and educational materials. Agencies should consider, as an initial step, using the Small Business Innovation Research (SBIR) program to solicit proposals for developing training materials and courses, to foster the uses of remote sensing data in applications, and to encourage commercial enterprises to provide these services.

Outreach

Finding. Reducing the social distance between application developers and end users is a means of encouraging successful technology transfer (see Chapters 2 and 3). Unless those who create applications (e.g., scientists, engineers, and technicians) and those who use them (e.g., government, not-for-profit, and private sector applied users, policy makers, and natural resource managers) understand the roles of others involved in the process, they will not be able to communicate effectively and the development of applications will suffer.

Recommendation 3. Federal agencies, including those that produce remote sensing images and those that use them, should consider creating "extern" programs with the purpose of fostering the exchange of staff among user and producer agencies for training purposes.

For example, NASA, NOAA, and USGS could create an extern program in collaboration with potential user agencies, such as the Environmental Protection Agency, the U.S. Army Corps of Engineers, the U.S. Department of Agriculture, the Department of Transportation, and others and in so doing could produce trained staff to serve as brokers for information and further training.

Similar exchanges could be organized between universities and state and local governments and between commercial companies and government.

Recommendation 4. The Land Grant, Sea Grant, and Agricultural Extension programs should be expanded to include graduate fellowships and associateships to permit students to work at agencies that use remote sensing data. Such programs could help to improve communication and understanding among the scientists and engineers who develop applications for remote sensing data and the agencies that use them.

NASA's Space Grant program could be extended to include these training activities, much as the Land Grant program has fostered the development of agricultural extension agents.

Applications Research

Finding. Although many remote sensing applications emerge from basic research, the development of applications is not accorded the recognition associated with publication in scientific journals. Researchers have few professional incentives to produce applications. The research-to-applications model developed in other fields, such as pharmaceutical research and many fields of engineering, could be emulated by the Earth sciences. Yet even if this model were to be adopted in areas related to remote sensing, there are at present few funding opportunities for work that spans the divide between research and applications.

Recommendation 5. Resources, separate from funding for basic research, should be made available to federal agencies such as NASA, the National Oceanic and Atmospheric Administration, the Environmental Protection Agency, the U.S. Geological Survey, the Department of Transportation, the National Science Foundation, and others for support of research on remote sensing applications and remote sensing applications derived from basic research. In addition, these agencies should establish joint research announcements aimed at fostering the development of applications for remote sensing data through basic research.

Requirements of Applications Users

Finding. Many remote sensing applications have specific requirements, including continuity in data collection, consistency in format, frequency of observations, and access to comparable data over time. It is important that the requirements of those who use applications are communicated to both public and private sector data producers throughout the process of designing new technologies and producing and disseminating remote sensing data.

Recommendation 6. Both public and private sector data providers should develop mechanisms to obtain regular advice and feedback on applications requirements for use in their planning processes. Advisory bodies that are consulted for input to these decisions should routinely include applications users.

Recommendation 7. Data preservation should be addressed by all data providers as a routine part of the data production process to ensure continuity of the data record and to avoid inadvertent loss of usable data.

Standards and Protocols

Finding. The lack of standard data formats, open and available protocols, and standard validation and verification information inhibits the spread of remote sensing applications (see Chapter 3).

Recommendation 8. The use of internationally recognized formats, standards, and protocols should be encouraged for remote sensing data and information. The work of the OpenGIS Consortium and the Federal Geographic Data Committee serves as an important international and national coordinating mechanism for efforts in standards development that should be continued.

These and other entities pursuing common remote sensing data formats and standards should consult with the sensor and software vendors to ensure that data acquired from the use of new technologies for data acquisition, analysis, and storage and distribution are consistent with other data sets.

Utility of Workshop Format

Finding. In general, the workshop as a mechanism for gathering data provided the steering committee with the information and insight it needed to understand issues related to technology transfer and remote sensing applications and to make recommendations about more effective ways to foster the development of applications.

In retrospect, as outlined in Chapter 4, the steering committee recognizes several strengths, and some areas for improvement, in the use of a workshop format.

B

Biographical Information for Steering Committee Members, Workshop Speakers, and Panelists

STEERING COMMITTEE MEMBERS

Roberta Balstad Miller, *Chair,* has worked and published extensively in the areas of science and technology policy and human interactions in global environmental change. She received her Ph.D. from the University of Minnesota. Currently the director of the Center for International Earth Science Information Network (CIESIN) at Columbia University, she was previously a staff associate with the Social Science Research Council (1975-1981), the founding executive director of the Consortium of Social Science Associations (COSSA) (1981-1984), and director of the Division of Social and Economic Science at the National Science Foundation (1984-1993). She received NSF's Meritorious Service Award in 1993. Dr. Miller has served as chair of a number of scientific advisory groups, including the NATO Advisory Panel on Advanced Science Institutes/ Advanced Research Workshops, the Committee on Science, Engineering and Public Policy of the American Association for the Advancement of Science, the Human Dominated Systems Directorate of the U.S. Man in the Biosphere Program, and others. From 1992 to 1994, she served as vice president of the International Social Science Council. Dr. Miller's NRC service includes former membership on the Climate Research Committee, the Global Change Research Committee, the Committee on the Geographic Foundation for Agenda 21, the Space Studies Board, the Board's Task Group on Research and Analysis Programs, and the Committee on Buildings and Community Systems Energy Conservation (Phase III).

Alexander F.H. Goetz has been a professor of geological sciences and director of the Center for the Study of Earth from Space (part of CIRES) at the University of Colorado, Boulder, since 1985. Dr. Goetz received degrees in physics, geology, and planetary science, all from the California Institute of Technology. Previously he spent 15 years at the NASA Jet Propulsion Laboratory, where he started and headed the geologic remote-sensing group and initiated the development of imaging spectrometry, now known as hyperspectral imaging. Prior to JPL he spent 3 years at Bellcomm, a subsidiary of AT&T Bell Labs, working on the Apollo program. Dr. Goetz has been a principal investigator in the Apollo, Skylab, shuttle, and Landsat programs. He was a member of the Landsat 7 science team and plays a similar role in the Earth Observing-1 (EO-1) satellite team. Dr. Goetz has received numerous awards, among them the NASA/Department of the Interior William T. Pecora award. In addition, Dr. Goetz was a founder and the CEO of Analytical Spectral Devices, Inc., in Boulder, for 10 years and is currently its chairman.

Lawrence W. Harding, Jr., is a research professor in the University of Maryland Center for Environmental Science, with appointments at Maryland Sea Grant and the Horn Point Laboratory. His research focuses on the use of aircraft and satellite remote sensing of ocean color to study phytoplankton responses to nutrient enrichment in estuarine and coastal waters. He also directs Sea Grant educational activities in remote sensing in collaboration with NASA scientists. Dr. Harding's main interests include coordination of a regional, multiplatform remote-sensing program in the Chesapeake Bay region to further the understanding of ecosystem health by applying new technologies to contemporary ecological issues.

John R. Jensen's research focuses on remote sensing of vegetation biophysical resources, especially inland and coastal wetlands; remote sensing of urban, suburban, and land use cover; the development of improved digital image processing classification, change detection, and error evaluation algorithms; and the development of educational materials for remote sensing instruction. Dr. Jensen has conducted contract and grant research for the Department of Energy's Savannah River Site, NASA commercial applications, and NOAA CoastWatch. He is the author of the remote sensing textbooks *Introductory Digital Image Processing: A Remote Sensing Perspective* and *Remote Sensing of the Environment: An Earth Resource Perspective*. He is the past president of the American Society for Photogrammetry and Remote Sensing (ASPRS) and currently serves on four National Research Council committees associated with remote sensing of the environment.

Chris J. Johannsen is director of the Laboratory for Applications of Remote Sensing (LARS) and professor of agronomy at Purdue University. His research interests are in spatial, spectral, and temporal aspects of remote sensing relating to geographic information systems (GIS) as applied to precision agriculture, land resource development, and land degradation. He was director of the Environmental Sciences and Engineering Institute (previously the Natural Resources Research Institute) (1988-1995) and director of the Agricultural Data Network (1985-1987) at Purdue University. From 1981 to 1985, Dr. Johannsen was the director of the Geographic Resources Center, Extension Division, at the University of Missouri at Columbia. Dr. Johannsen has been named a fellow of the American Society of Agronomy, the Soil Science Society of America, and the Soil Conservation Society of America and is a member of the International Soil Society, the American Society of Photogrammetry, and Sigma Xi. He has served on the Space Studies Board's Committee on Earth Studies (1995-1998), the Committee on NASA Information Systems (1986-1987), and the Panel on Earth Resources (1982-1983).

Molly Macauley is a senior fellow at Resources for the Future (RFF), where she directs its space economics research program. Her research interests include space economics and policy; recycling and solid waste management; urban transportation policy; and the use of economic incentives in environmental regulation. An economist at RFF since 1983 and a long-time analyst of the commercial use of space technology, Dr. Macauley offered her views to Congress in May 1997 on how government can foster burgeoning commercial ventures into satellite remote sensing. One of her major research projects looks at the ongoing economic—as well as privacy, security, and other—implications of American companies selling images photographed by privately owned satellites in outer space. Her other research projects are exploring the use of economic incentives to manage space debris; the allocation of scarce energy, water, utilities, and telecommunications resources on the ISS; the value of geostationary orbit; and the value of information, particularly information derived from space-based remote sensing. She was a member of the Space Studies Board's Task Group on Setting Priorities for Space Research and the National Research Council (NRC) Committee on the Impact of Selling the Federal Helium Reserve. She also served on the NRC's Aeronautics and Space Engineering Board's Committee for the Assessment of the National Aeronautics and Space Administration's Space Solar Power Investment Strategy.

John S. MacDonald is a consultant and chair of the Institute for Pacific Ocean Science and Technology. He is one of the founders of MacDonald, Dettwiler and Associates (MDA) Ltd. Dr. MacDonald was responsible for all aspects of business operations, overall strategic leadership, technical leadership, and market positioning worldwide. His interests lie in the areas of advanced digital systems engineering, remote sensing, and image processing. He led the design team for the first Landsat ground-processing system produced by MDA, Ltd., and was involved in the early development of synthetic aperture radar processing at this company. His technical activities have been in the areas of information extraction from advanced sensor systems and the applications of remote sensing, with particular emphasis on the physics of the backscatter process and the use of integrated data sets as a means of increasing the ability to extract useful information from remotely sensed data.

Jay S. Pearlman is development team manager at TRW, Inc. His background includes basic research, program management, and program development in sensors and systems. He has played an important role in the development and implementation of new concepts and capabilities for both the military and the civil sectors of the U.S. government. He is currently working on the EO-1 Hyperion sensor as principal investigator and is actively involved with the EO-1 Science Validation Team in assessing the benefits of hyperspectral imagery. He is also involved in an assessment of the viability of multispectral and hyperspectral commercial applications.

WORKSHOP SPEAKERS AND PANELISTS

G. Bryan Bailey is principal remote sensing scientist at the U.S. Geological Survey's EROS Data Center (EDC), near Sioux Falls, South Dakota. He holds B.A. and M.S. degrees in geology from the University of Iowa and a Ph.D. in mineral exploration from Stanford University. Prior to joining the EROS program in 1978, Dr. Bailey spent 4 years with ASARCO, Inc., conducting base and precious metal exploration in the western United States. During his career at the EDC, Dr. Bailey has been responsible for conducting and directing research and applications development in the field of geologic remote sensing, and he has been involved extensively in program development and implementation at EDC, particularly as it relates to USGS and EDC involvement in the Landsat and Earth Observing System (EOS) programs. As EDC Distributed Active Archive Center (DAAC) project scientist, he was responsible for liaison between the EDC DAAC and the EOS and global change science communities throughout the development, implementation, and operation of DAAC systems and capabilities. He continues to provide scientific and programmatic liaison between the EDC DAAC and the Advanced Spaceborne Thermal Emission and Reflection Radiometer (ASTER) science team and project, including participating as validation scientist for two ASTER DEM standard data product validation sites. Currently, Dr. Bailey is responsible for initiating and leading USGS activities designed to expand and enhance beneficial use of remotely sensed data throughout the USGS. He is the author of many scientific publications and other program- and policy-related papers.

Donna Boreck works as a mitigation specialist with the Federal Emergency Management Agency (FEMA). Prior to coming to FEMA, she worked as a research scientist on geologic and environmental issues involved in development of energy and mineral resources. More recent work has addressed drinking water protection and assessing health risks. She has a bachelor's degree in geology from Colorado State University and a master's degree in geology from the Colorado School of Mines. She has published over 20 papers in her field.

Mark Bosworth has 10 years of active work experience in geographic information systems (GIS) technology. After graduating from Hunter College in New York with a bachelor's degree in cartography and a master's degree in analytical cartography and remote sensing, Mr. Bosworth moved to Portland, Oregon, and joined the Data Resource Center (DRC) at Metro as a GIS specialist. As a GIS program supervisor at Metro, he manages all of the Data Resource Center's public interfaces, including development of an electronic GIS storefront on Metro's Web site (www.metro-region.org), the DRC's map counter services, and ongoing development of Metro map products. Mr. Bosworth also serves as Metro's GIS liaison with state and national transportation agencies. He has made presentations to local, regional, state, and national GIS audiences on regional strategies for public participation in GIS, sharing of data on roads, generic linear reference systems, and Metro's Regional Land Information System. He has published technical papers on address geocoding, systems administration in GIS, and road data sharing strategies. He is particularly interested in how public participation in GIS can be enhanced and what the impact will be on cartographic data structures and access policies.

Patrick Bresnahan is currently the geographic information officer for Richland County, South Carolina. He is an American Society for Photogrammetry and Remote Sensing (ASPRS) certified mapping scientist in GIS/land information systems and a member of NASA's Program Planning and Analysis Panel. Currently, he co-chairs the technology subcommittee of the South Carolina State Mapping Advisory Committee. He earned a bachelor's degree (University of Maryland, Baltimore County), a master's degree (Indiana State University), and a Ph.D. (University of South Carolina) in geography, and he participated in the Postgraduate Research Program for the U.S. Department of Energy (DOE) and later was awarded a postdoctoral research fellowship sponsored by the Oak Ridge Institute for Science and Education. His postdoctoral research was conducted at the DOE Savannah River facility. Dr. Bresnahan also received fellowship awards from Automated Mapping and Facilities Management International (now Geospatial Information and Technology Association (GITA)) and NASA under the South Carolina Space Grant Consortium. He maintains membership in ASPRS, the Association of American Geographers, GITA, and Sigma Xi and has remained active in those organizations through numerous conference presentations and publication of research results in *Photogrammetric Engineering and Remote Sensing*.

Andrew J. Bruzewicz is director of the U.S. Army Corps of Engineers' Remote Sensing/GIS Center, Hanover, New Hampshire, and acting associate technical director for Geospatial Research and Development. He manages the Corps's civil works geospatial research and development program area (survey and mapping, remote sensing, and GIS) and is currently involved in the creation of new approaches to the development and life-cycle support of science and engineering technology products. His primary professional interests are the integration of remote sensing and GIS into the Corps's mission areas, with particular emphasis on emergency management, data sharing within and between agencies, and remote sensing and GIS education.

Peter Conrad is a GIS coordinator/environmental planner for the Baltimore City Department of Planning. He holds a bachelor's degree in political science with an environmental studies minor from the University of Florida and a master's degree in urban and regional planning from the University of Wisconsin. He joined the Baltimore City Department of Planning in 1987, working as a community planner from 1987 to 1993. He became an environmental planner in 1993 and began work with GIS in 1998. He is coordinating several GIS projects, including conflation of 2000 Census geography, analysis of Baltimore's neighborhood real estate market, and development of a multiattribute land use map for mapping forest areas. Mr. Conrad is also assisting the Maryland Department of Natural Resources in using IKONOS imagery to map natural land cover, including forests, in Baltimore City. These data will be used for a variety of purposes, including a citywide natural resource inventory, planning for urban reforestation, and the development of a land use/land cover map for Baltimore.

John Dorman is currently assistant director for emergency management—hazards mapping, information, and management. His work involves floodplain mapping, floodplain management, flood warning, development and management of geographic information systems that support Hazard US (HAZUS) development and mapping, and antiterrorism planning and mitigation, among other responsibilities. Previously, he was administrator for planning in the North Carolina Office of State Budget, Planning and Management, where he was responsible for statewide strategic planning and evaluation; sociodemographic data and demography; the Center for Geographic Information and Analysis; the North Carolina Geodetic Survey; and the North Carolina Floodplain Mapping Program. Former positions also included deputy state planning director, North Carolina Office of State Planning. In June 2001, the state of North Carolina, through the North Carolina Floodplain Mapping Program, received the Tom Lee State Award for Excellence–Platinum Level from the Association of State Floodplain Managers. This award is given annually to recognize an outstanding floodplain management program or activity at the state level.

David Ekern is associate director of the American Association of State Highway and Transportation Officials and assistant commissioner of the Minnesota Department of Transportation. He is currently on assignment to the American Association of State Highway and Transportation Officials, focusing on initiatives that are changing the face of our nation's transportation agencies. Focus areas include intelligent transportation systems, asset management, remote sensing technologies, operations management, and context-sensitive design. At the Minnesota Department of Transportation, he has served as assistant commissioner for national and international programs, division director of engineering services, and assistant chief engineer, and as a district engineer. He also has held positions in environmental policy and planning, preliminary design, MPO and regional planning, and highway maintenance. He is a member of numerous professional associations and societies and is a registered professional engineer. Mr. Ekern received a B.S. in civil engineering from the University of Minnesota and an M.B.A. from the University of St. Thomas.

Joe Engeln is the assistant director for science and technology at the Missouri Department of Natural Resources. In this position, he advises the director on a broad range of interdisciplinary environmental issues and on data management. The department is the delegated state agency for environmental quality and operates the state parks, historical sites, and soils program. Prior to joining the department in 1999, he was an associate professor of geological sciences at the University of Missouri. From 1990 to 1992, Dr. Engeln was the acting geodynamics program scientist at NASA Headquarters. He has written papers on seismology, tectonics, hydrology, geodesy, marine geophysics, geodynamics, and remote sensing.

Amanda Hargis, currently the GIS coordinator for Boulder County, Colorado, has been working with GIS since 1989. In her enthusiasm for GIS, she has been involved in organizing local and national GIS conferences for several years. She also facilitates the Boulder Area Spatial Information Co-op (BASIC), which is a regional group of agencies that share digital information such as parcels, air photos, and other imagery.

Michael Hove is data processing coordinator for the North Dakota State Water Commission, where he has worked for 25 years. He has worked on five projects with lidar data over the last 3 years, using lidar to collect detailed topographic data and help understand overland water flow. The primary focus of the Water Commission is ground water and surface water resource management. Mr. Hove holds a bachelor's degree in computer science.

Charles Hutchinson is acting director of the Division of Applications for NASA's Office of Earth Science. Dr. Hutchinson is a professor in the Arid Lands

Resources Sciences Program at the University of Arizona, where he also serves as director of the Arizona Remote Sensing Center and associate director of the Office of Arid Lands Studies in the College of Agriculture. Previously, Dr. Hutchinson held positions at the Jet Propulsion Laboratory and the USGS Earth Resources Observation Systems (EROS) Program Office. His research has focused on the use of remote-sensing and geographic information system technology for monitoring environmental change and food security issues in arid lands, focusing largely on conditions in the American Southwest and in Africa. Dr. Hutchinson serves as the executive editor for the *Journal of Arid Environments* and is executive secretary for the International Committee on Remote Sensing of the Environment, International Society for Photogrammetry and Remote Sensing. He holds a Ph.D. in geography from the University of California at Riverside.

Jeff Liedtke has been actively involved in the geotechnology industry for more than 15 years. As director of market planning for DigitalGlobe, Inc., he is responsible for identifying and leveraging core technology and partnerships to develop products and markets. Prior to that, he was the product marketing manager at Space Imaging, Inc., for IKONOS, Landsat, Indian Remote Sensing, and value-added products. He provided business development, including international project management, development of remote sensing and digital photogrammetry software and hardware systems, and sales and marketing for International Imaging Systems. Previously, he also worked for the Los Padres National Forest to develop a GIS database in mapping, hydrology, soils, geology, and fire management, producing maps for all land management prescription scenarios. Mr. Liedtke holds B.A. degrees in geography and environmental studies from the University of California at Santa Barbara, and an M.S. in remote sensing from Simon Fraser University, B.C., Canada.

Don Light is a senior consultant for mapping and remote sensing companies. Previously, he was manager of business development at Emerge and manager of business development for commercial remote sensing at Eastman Kodak. Mr. Light's background includes a number of professional positions in defense and civilian mapping and private sector companies. Mr. Light served in the 30th Engineer Topographic Battalion from 1953 to 1955, was a member of NASA's Apollo Photo Team and the Large Format Camera Team, and served as chief of the Advanced Technology Division at the Defense Mapping Agency (DMA) Topographic Center and as the technical director of the Defense Mapping School. After an assignment at the Office of Technology Assessment, where he worked on a study of National Space Policy and Remote Sensing Applications, he transferred to the U.S. Geological Survey's (USGS's) National Mapping Division as chief of the Branch of Systems Development. In 1987, he became chief, Office of Production Contract Management, where he was responsible for the National

Aerial Photography Program, the Airborne Radar Imagery Program, initiation of the Digital Orthophoto Program, and the National Camera Calibration Laboratory. In 1991, he was awarded the Department of the Interior Meritorious Service Citation for excellence in managing the USGS imaging programs. He has a long affiliation with the American Society of Photogrammetry and Remote Sensing (ASPRS), including serving as director of the Primary Data Acquisition Division, and also served as co-chair of the International Society of Photogrammetry and Remote Sensing (ISPRS) Working Group on Mapping from High Resolution Imagery, and is a member of the ISPRS editorial board. Mr. Light became an ASPRS Photogrammetric Fellow in 2000. He holds a B.S. in geodetic and cartographic science from George Washington University, as well as a graduate diploma in strategy and management from the U.S. Naval War College and a diploma from the Federal Executive Institute.

Michael Renslow is the vice president of Spencer B. Gross, Inc. (SBG). He has 34 years of experience in the mapping sciences as an engineering surveyor, professional cartographer, photogrammetrist, aerial photographer, and business manager. He worked for the Department of Defense, the U.S. Forest Service, and two prominent aerial photography firms before joining SBG in 1995. He is a 1971 graduate of San Francisco State University with a B.S. in geography and training in civil engineering and geology. Mr. Renslow serves as an adjunct faculty member at the University of Oregon Geography Department, teaching courses in remote sensing. He has recognized expertise in lidar (light detection and ranging) mapping technology and is currently conducting research to characterize and measure forest biometrics and wildlife habitat zones.

Rebecca Storey is a contractor with the Denver Regional Office of the Federal Emergency Management Agency (FEMA). Her professional work with FEMA began in 1994 with the January 1994 Northridge Earthquake in California, for which she supported the Human Services program area by processing federal assistance to disaster victims. She has also served as disaster reservist in Aberdeen, South Dakota, where she prepared GIS mapping of roads inundated in a flood disaster, and as the technical services branch chief during the Susquehanna River flood in Harrisburg, Pennsylvania. In addition, Ms. Storey produced mapping with georeferenced sites of dams and culverts connected to major rivers in several flooded North Dakota counties for the FEMA Regional Office in Denver, and in January 1997 she served as a co-GIS coordinator for flooding in Salem, Oregon, before being deployed to North Dakota in response to the Grand Forks flood of record to produce GIS for disaster management staff. Prior to joining FEMA, Ms. Storey served as a city planner for 4 years.

David Thibault is executive vice president of Earth Satellite Corporation (EarthSat). He has been a member of the technical and management staff since

1971 and provides leadership for financial and long-range planning, as well as business development. Mr. Thibault's broad background in public administration and policy making at the state and federal levels of government has been expanded at EarthSat on projects designed to apply remote sensing technology to operational programs and long-range planning. Mr. Thibault served as director of research in the Massachusetts governor's office and on a White House task force directed at improving federal-state relations. At the U.S. Department of Commerce from 1968 to 1971, he served as special assistant in the Office of the Secretary, where he worked on the departmental programs, including regional economic development. Mr. Thibault has had responsibility for many of EarthSat's state and local projects.

Melanie Wallendorf is Eller Professor of Marketing in the College of Business and Public Administration at the University of Arizona. She received an M.A. in sociology in 1977 and a Ph.D. in marketing in 1979, both from the University of Pittsburgh. Her research focuses on sociological aspects of consumer behavior. In particular, Dr. Wallendorf's research has made pioneering contributions to the study of consumption using ethnographic research methods. For over two decades, she has published extensively on the sociocultural meanings of consumption, including the role of possessions in socially constituting social classes and ethnic groups, the meanings of favorite possessions and of collections, the processes that define particular possessions as sacred or profane, and the impact of social ties on the diffusion of innovations. In 1992 she won the Association for Consumer Research Award for the best article in the *Journal of Consumer Research* 1989-1991 for her article "The Sacred and the Profane in Consumer Behavior: Theodicy on the Odyssey" (co-authored with Russell Belk and John F. Sherry, Jr.).

Robert A. Wright is the assistant information technology division manager for GIS for the Washington Department of Natural Resources. Previous positions include principal consultant for R.A. Wright and Associates, a remote-sensing marketing group, and director of reseller marketing and director of product marketing, providing worldwide marketing management for EarthWatch, Inc.'s. QuickBird satellite imagery products. In addition, Mr. Wright has also provided system integration, consulting, marketing, and remote sensing/GIS support for Atterbury Consultants, Inc., and served as the director of NW Operations for Computer Sciences Corporation, providing GIS systems integration and related services in support of the company's GIS Center of Excellence and on several major contracts. Mr. Wright served as the land information systems manager for Infotec Development, Inc., and directed the development of its LIS/GIS business area. In addition, he has served as the state office GIS manager for the Oregon State Office of the Bureau of Land Management; as GIS coordinator for the Portland Area Office of the Bureau of Indian Affairs; and as area forester and agency forester in California and New Mexico locations over a 25-year career with the Department of the Interior. He graduated with a B.S. in forest management from Oregon State University. He is a member of the Society of American Foresters and of ASPRS and was secretary-treasurer, vice president, and president of the Columbia River Chapter of ASPRS, 1996-1998; national director, 1999; treasurer of the GIS in Action conference, ASPRS 1999; National Conference assistant director; Rocky Mountain region secretary and vice president, 2000-2001; Puget Sound Region vice president, 2001; and president elect, 2002.

Jeff Young, executive director for Global Solutions Sales Space at Imaging, Inc., is responsible for sales of Earth imagery and value-added information products, services, and solutions. He has more than 25 years of experience in sales, as well as program and project experience, including more than 12 years in senior management of GIS corporations. Before joining Space Imaging, Mr. Young worked for Bentley Systems, Inc., as an account manager for the Government and Transportation programs. Prior to that, he held numerous professional positions in the GIS and remote-sensing field, including president of the Criminal Justice Business Unit at Graphic Data Systems; project manager and senior GIS consultant for the Program Administration Group; GIS consultant with HDR, Inc.; and scientific supervisor and principal investigator for Lockheed Engineering and Management Services Company. Mr. Young holds an M.A. degree in geography from Arizona State University and a B.S. in geography from Lock Haven University.

C

Workshop Agenda and Participants

8:00 a.m. **Continental Breakfast**

8:30 **Welcome**
Prof. Susan Avery, Director, University of Colorado, Cooperative Institute for Research in Environmental Sciences (CIRES)

8:35 **Opening and Key Questions for the Workshop**
Roberta Balstad Miller, Chair, Steering Committee on Space Applications and Commercialization

8:45 **Panel I: Opportunities and Impediments in Using Remote Sensing Data for State and Local Governments**
Moderator: *John Jensen, Steering Committee member*

Speakers:
Patrick Bresnahan, Richland County, South Carolina
Joe Engeln, Missouri Department of Natural Resources
Peter Conrad, Baltimore City Department of Planning
Amanda Hargis, Boulder County GIS Coordinator

This session will explore issues regarding the use or potential use of remote sensing data from a public sector perspective. Topics such as data access, cost, technical skills for remote sensing use, data networks, sharing and reporting, and how federal regulations affect public sector data requirements may be discussed.

10:30   Break

**11:00   Panel II: Doing Business with State and Local Governments:
                The Private Sector Perspective**
Moderator: *John MacDonald, Steering Committee member*

Speakers:
Jeff Young, Space Imaging, Inc.
Jeff Liedtke, DigitalGlobe, Inc.
David Thibault, Earthsat, Inc.
Don Light, PAR Government Systems

- How difficult or easy is it to do business with states, as compared to other clients?
- Why are states using remote sensing data for some applications and not others? And why have some states adopted remote sensing technology while others have not?
- What are the particular requirements or issues of working with states and local governments, and how have they been addressed?
- What opportunities do you see for better inserting remote sensing data into state applications or processes?

12:30 p.m. Lunch

1:45   Applications Theme I: Emergency/Disaster Response
Moderator: *Alex Goetz, Steering Committee member*

Case Study: The North Dakota Red River Valley Flood of 1997
Michael Hove, North Dakota State Water Commission

Case Study: Responding to Hurricane Floyd: The North Carolina
    Floodplain Mapping Program
John Dorman, North Carolina Floodplain Mapping Program

Overview of Remote Sensing and Emergency/Disaster Response
Andrew Bruzewicz, U.S. Army Corps of Engineers

General Discussion

3:15   **Break**

3:45   **Keynote I: Emerging Application Area: Transportation**
Moderator: *Molly Macauley, Steering Committee member*

David Ekern, Association of State Highway Transportation Officials/ Minnesota Dept. of Transportation/Chair of NRC Transportation Research Board Remote Sensing Committee

4:30   **Keynote II: Adaptors and Adopters of Remote Sensing Data**
Moderator: *Chris Johannsen, Steering Committee member*

Speaker: *Melanie Wallendorf, University of Arizona, Department of Marketing*

5:15   **Adjourn**

Thursday, January 24, 2002

8:00 a.m.   **Continental Breakfast**

8:30   **Keynote III: The Use of Lidar in State and Local Applications**
Moderator: *Chris Johannsen, Steering Committee member*

Michael Renslow, Spencer B. Gross, Inc.
- Why is lidar being adopted more readily than satellite-based remote sensing data?
- What makes lidar appealing and worthwhile?
- What was the impetus for its applications at the state level and for the growing use of the technology?

9:15   **Applications Theme II: Land Use/Planning**
Moderator: *Jay Pearlman, Steering Committee member*

Case Study: Washington State Land Trust/Management
Robert Wright, Washington State Department of Natural Resources

Case Study: Monitoring Regional Land Use/Land Planning in the Portland, Oregon, Regional Area
Mark Bosworth, Metro (Portland Regional Planning Organization)

Overview of Remote Sensing and Land Use/Land Planning:
Patrick Bresnahan, Richland County, South Carolina Department of Information Technology

General Discussion

10:45   Break

11:00   Perspectives on Federal Support for Public Sector Applications
Moderator: *Roberta Balstad Miller, Steering Committee chair*

Speakers:
G. Bryan Bailey, USGS
Charles Hutchinson, NASA Headquarters
Rebecca Storey/Donna Boreck, FEMA Region 8, Denver

This session will explore resources and services available to state and local governments, such as technical support, training, and applications products. Speakers will address the services and support offered by their institutions to state, local, city, county, or regional governments and the rationale behind this support.

12:30 p.m. Wrap-up and Adjourn
Roberta Balstad Miller, Chair

WORKSHOP PARTICIPANTS

Avery, Susan, CIRES, University of Colorado
Bailey, G. Bryan, U.S. Geological Survey
Bailey, G. Paul, Environmental Systems Research Institute, Inc.
Boreck, Donna L., Federal Emergency Management Agency
Bosworth, Mark, Portland, Oregon, METRO
Brender, Mark, Space Imaging, Inc.
Bresnahan, Patrick J., Richland County, South Carolina, Department of
    Information Technology
Brett, Thomassie, DigitalGlobe, Inc.
Bruzewicz, Andrew J., U.S. Army Corps of Engineers
Cary, Tina, Cary and Associates
Cholvin, Brooke, Boulder County Assessor's Office
Conrad, Peter, Baltimore City Department of Planning
Correa, Aderbal C., University of Missouri at Columbia
David, Leonard, Space.com
Davis, Ron, DigitalGlobe, Inc.

Dorman, John K., North Carolina Floodplain Mapping Program
Driese, Kenneth L., WyGISC
Eadie, Rob, 3Di Technologies, Inc.
Ekern, David S., American Association of State Highway and Transportation
     Officials
Engel-Cox, Jill, Battelle Memorial Institute
Engeln, Joe, Missouri Department of Natural Resources
Fisher, Verlin, 3Di Technologies, Inc.
Friedl, Lawrence, U.S. Environmental Protection Agency
Goetz, Alexander F.H., CIRES, University of Colorado
Gwaltney, Gregory, U.S. Environmental Protection Agency
Harding, Lawrence W., Horn Point Laboratory, University of Maryland
Hargis, Amanda, Boulder County GIS
Hastings, David, NOAA National Geophysical Data Center
Hernandez, Michael W., University of Missouri at Columbia
Hoff, Raymond M., University of Maryland, Baltimore County
Hori, Kenta, Hitachi Software Global Technology
Hove, Michael, North Dakota State Water Commission
Hutchinson, Charles, NASA Headquarters
Irani, Frederick M., Maryland Department of Natural Resources
Jensen, John R., University of South Carolina
Johannsen, Chris J., Purdue University
Johnson, Nan, City of Boulder Planning Department
Kagan, Jimmy, Oregon Natural Heritage Program
Leung, Ada, University of Arizona
Liedtke, Jeff, DigitalGlobe, Inc.
Light, Don, Consultant, Airborne and Space Systems
Macauley, Molly, Resources for the Future
MacDonald, John S., Institute for Pacific Ocean Science and Technology
McCraw, David J., New Mexico Bureau of Geology and Mineral Resources
McCuistion, Doug, NASA Goddard Space Flight Center
McFadden, Bryan, Spacing Imaging, Inc.
Miller, Roberta, Columbia University
Napier, Gary, Space Imaging, Inc.
Palmerlee, Thomas, National Research Council Transportation Research Board
Pardue, Jaymes, EMERGE
Pearlman, Jay, TRW, Inc.
Pratt, Kristina, Boulder County Parks and Open Space
Price, Kevin, University of Kansas
Raber, Steve, NOAA Coastal Services Center
Renslow, Michael, Spencer B. Gross, Inc.
St. Pierre, Marcel, Canadian Space Agency
Sherry, Jennifer, City of Boulder National Parks

Storey, Rebecca A., Federal Emergency Management Agency
Tessar, Paul, Colorado Department of Transportation
Thibault, David, EarthSat, Inc.
Thirumalai, K., U.S. Department of Transportation
Thomas, Michael, NASA Stennis Space Center
Tilley, Jay, Resource 21
Tuohy, Mark, Space Imaging, Inc.
Vogel, David, DigitalGlobe, Inc.
Wallendorf, Melanie, University of Arizona
Warnecke, Lisa, GeoManagement Associates
Williams, David J., U.S. Environmental Protection Agency
Wollack, Leslie, National States Geographic Information Council
Wright, Robert, Washington Department of Natural Resources
Yoke, Lyne, National Snow and Ice Data Center
Young, Doug, District Office of Rep. Mark Udall
Young, Jeffrey M., Space Imaging, Inc.

D

Acronyms

ASPRS   American Society for Photogrammetry and Remote Sensing
AVHRR   Advanced Very High Resolution Radiometer

BAA   Broad Agency [Research] Announcement
BASIC   Boulder Area Spatial Information Cooperative

DEM   digital elevation model
DFIRM   Digital Flood Insurance Rate Maps
DNR   Department of Natural Resources

EOSDIS   Earth Observing System Data and Information System
EROS   Earth Resources Observation Systems

FEMA   Federal Emergency Management Agency
FOIA   Freedom of Information Act

GIS   geographic information system
GPS   Global Positioning System

IRS   Indian Remote Sensing Satellite

lidar   light detection and ranging

MODIS   Moderate-resolution Imaging Spectroradiometer

NASA	National Aeronautics and Space Administration
NCDC	National Climate Data Center
NESDIS	National Environmental Satellite, Data, and Information Service
NOAA	National Oceanic and Atmospheric Administration
NPOESS	National Polar-orbiting Operational Environmental Satellite System
NRC	National Research Council
RESAC	Regional Earth Science Applications Center
RFP	request for proposal
RLIS	Regional Land Information System
SPOT	Système pour l'Observation de la Terre
USGS	United States Geological Survey
WAGIC	Washington State Geographic Information Council